Motorways versus Democracy

Public inquiries into road proposals and their political significance

John Tyme

First published 1978 by
THE MACMILLAN PRESS LTD
London and Basingstoke
Associated companies in Delhi Dublin
Hong Kong Johannesburg Lagos Melbourne
New York Singapore and Tokyo

Printed in Great Britain by
REDWOOD BURN LTD
Trowbridge and Esher

British Library Cataloguing in Publication Data

Tyme, John
Motorways versus democracy.
1. Highway planning – Great Britain – Citizen participation 2. Express highways – Great Britain – Planning
I. Title
711'.73'0941 HE363.G73C/

ISBN 0–333–23187–2
ISBN 0–333–23188–0 Pbk

Contents

Chronology of Inquiries

1973		
June/July	M42	Bromsgrove
September	M40	Kenilworth
October	M20	Ashford (Kent)
1974		
February	M56	Chester
February/March	M65	Burnley
December	M16	Epping
1975		
January	M16	Epping
May	M25	Byfleet
May	A55	Llandudno
November	Aire Valley Trunk Road	Shipley
1976		
January	Elmstead By-pass	Colchester
January	Hayle By-pass	Penzance
February	Lewes By-pass	Lewes
April	Haslingden By-pass	Haslingden
May	M25	Hornchurch
June/July	M3	Winchester
September	Archway Road I	Hornsey
September	M27	Chichester
1977		
April	Archway Road II	Archway
May	Ipswich By-pass	Ipswich
October	Archway Road III	Archway
November	M25	Ripley

The author and publishers would like to thank Mike Peyton, whose perceptive and entertaining cartoons appear in this book.

Foreword

by David Widdicombe, Q. C.

One does not have to agree with everything John Tyme has said and done to recognise the important service which he and his supporters have performed in opening up to scrutiny and debate the whole subject of decision-making in the field of Government-sponsored projects. Our public inquiry system works well enough when the proposal under investigation is promoted by a local authority or private developer – that is, someone other than the Government. But as applied to Government-sponsored projects like motorways it is thoroughly unsatisfactory. This is partly because the Government combines the roles of advocate for the scheme and judge of the objections to it, compounding this unfairness by appointing as inspectors at the inquiries persons such as ex-civil servants who do not strike the public as independent. Far more important, though, is the embargo placed at the inquiry on all discussion on the main thing everyone wants to discuss, namely whether the project should take place at all.

The plain fact is that people everywhere in this country today want more say in the decisions that affect them, whether at their place of work or where they live. Voting once every few years may have been enough a hundred years ago, but it is not enough for a community which has enjoyed the benefits of universal education for several generations. Our institutions must develop to accommodate this demand for participation. In particular, this book shows that we must devise ways to involve people in decision-making in the field of transport policy. If policy cannot reasonably be discussed at the local inquiry, new institutions must be set up at which the need for a project can be fairly and fully debated. Parliament has not got the time to discuss such matters, and is in any event hampered by the polarisation of opinion which occurs at the political level. It may be that we need a range of institutions lying between the public local inquiry and Parliament at which policy can be debated. The other feature essential for participation is, of course, open government – an idea to which the Department of Transport is notoriously resistant.

John Tyme has by his persistent campaign reflected and channelled the general dissatisfaction with the decision-making process in the transport field, which has developed as public awareness of environmental issues has come to the fore in recent years. I particularly commend to the attention of readers Chapter 7, in which he describes the part played in decision-making by interest groups, who can sometimes establish disquietingly close relationships with Government departments at official level. John Tyme is quite correct, in my view, to identify and bring into the open this element in the process.

It is to the credit of our system of government that the body of opinion which John Tyme represents has not been brushed aside. Its importance has been accepted, and his work is bearing fruit. Rules have now been laid down for motorway inquiries. A committee has been set up to investigate procedure. The Leitch Committee has reported on the Department's methods of traffic forecasting and appraising schemes, making recommendations which go a long way to justifying John Tyme's criticisms, and calling, as he does, for more openness. Other reports will follow. There is still a lot to do, but John Tyme has set the process of reform well and truly in motion.

March 1978 D. W.

Preface

In this short volume I have sought to do a number of things. First to describe how I came to be involved in inquiries into road schemes in the first place, and then in something of a narrative form to give an account of the inquiries themselves. In some instances I have included submissions in full in the text as any curtailment would lead to incomplete understanding of the issues involved, while others I have summarised in the text and set some down in full in the appendices. Press comments have been included when I have found them both to be accurate and to convey the flavour of events. Finally I have sought to draw some fundamental conclusions from it all, and the book conveniently ends with the implications of a High Court case concerning the same inquiry that started everything off.

Inevitably it tends to be a personalised account, and I much regret that space does not allow adequate mention of the work of the many highly intelligent and dedicated people with whom I have been associated, and who so often took up where I left off. The book, then, seeks to be no more than the account of the intervention of an ordinary non-expert person in a field of central government planning involving expertise, power and authority of the highest order. It is an account of how this enormous – and corrupted – power has been brought low, not by counter-expertise, purchased at great expense, but by quite ordinary people from all walks of life and from the whole breadth of the political spectrum – quite ordinary people asking quite ordinary questions, requesting that quite ordinary measures should be taken to provide the answers and, when denied those, taking the necessary extraordinary and courageous action.

That the writer was invariably at the centre of the many controversial events that followed makes the personalisation of this account inevitable. This may lead to accusations of a personality cult or whatever, and if this pleases some, then no harm is done. What would please me would be for this book to be read by people who, overwhelmed by bureaucratic power and obduracy of one form or another and denied the protection they have a right to expect, are thus enabled to see that by entirely peaceful means and drawing on an ancient and honourable tradition in the history of this country – that of Dissent – they can achieve justice and defend themselves and their communities not by playing the game according to the rules which (in so far as they are revealed) are devised by their opponents, but by working out a new game altogether, with rules devised by themselves.

For it is my conviction that so long as institutional protection is absent only ordinary, common-sense people banded together, people untainted by the sick dreams of the great technocrats, can halt the drive towards social and political disaster. First by asking simple questions like: Do we need it? Is it sensible, humane? What real purpose does it serve? And then by an evolutionary process asking further questions: Is it in accord with government policy? What is that policy anyway? Have the proposers fulfilled their statutory and constitutional obligations? And thence to questions beyond: What is happening to our democracy and our parliamentary system of government? Where are we heading? And finally, judging from the answers they receive or work out for themselves, taking various courses of action within the scope of that alluded to above. It is to them, therefore, that this book is primarily directed, and it is dedicated, of course, to those courageous friends and acquaintances that I have made throughout the events described in these chapters. Its dedication therefore writes itself.

To all those who have had the courage
to stand up (or sit down on the floor)
and be counted, who by their actions
constitute a bright candle in an
otherwise darkening world

Prelude

Neither in equity nor in law has the Minister the power to deny grounds of objection to the public.

Counsel for the M.M.A.C. at the M42 inquiry

This book is concerned not with roads themselves, but with events, many of them not without drama, that have occurred within certain road inquiries, and with the legal and constitutional issues which gave rise to them. But a question may be asked: Why the opposition to roads and motorways anyway, why all the fuss about them in particular?

As is made clear in a forthcoming book,* it is my belief, and one shared by increasing numbers of people, that the motorway/trunk road programme with all its ramifications poses a consummate evil, and constitutes the greatest threat to the interests of this nation in all its history. None of our national enemies have so mutilated our cities, undermined the long-term economic movement of people and goods, destroyed our industrial base, diminished our ability to plan our community life, and reduced our capacity to feed ourselves. The more highways we build, the more we generate traffic to fill them, the greater the congestion and snarl-ups, and thus the more highways we require to build. The more we build, the more we confirm and perpetuate the horrendous accident level (approaching a million people a century killed, to say nothing of the mutilated and injured) as motorway-generated traffic makes its way onto crowded city and suburban streets. The more roads and motorways built, the more inevitable is the decline of alternative transport modes. The more roads, the greater the housing loss and destruction of community and

* John Tyme, *Roads to Ruin* (London: Macmillan, forthcoming).

the less house-building and resources for hospitals, schools and other social services. The more highways, the more we are committed to the disaster known as 'dispersal planning' based upon the notion that distance between residence and work, shops and schools, recreation and medical services is no object; and the more dispersal planning, the greater the loss of land and agricultural production (now estimated at an average county area every ten years). The more resources we commit to road transport, the more we create social inequity (with all its imponderable political dangers), as the well over 40 per cent of households who do not own cars and are now never likely to, are left unable to pay the rising cost of public transport, simply watching the cars and juggernauts go by. The more we construct highways, the more we fuel the inflationary spiral as people are compelled to buy and maintain cars they cannot afford simply in order to get to work or get their children to school, to the dentist, the doctor or the hospital. The more motorways, the greater is our national dependence on the car industry, the one industry that, for reasons of energy and materials costs, can have no medium- let alone long-term future; the more roads, the greater the threat of unemployment of nightmare proportions as that industry and all its associated industries collapse before a vanishing world market. The more roads planned, the greater the industrial as well as housing blight, as blue, orange, green and red routes lie across our city maps for decades. The more the concrete miles proliferate, particularly in development areas, the more economic decline proceeds as direct investment declines in industry and in housing and those social services which together stimulate economic activity and create a contented work force. The more freight and personal movement we commit to roads, so the self-proliferating highway programme can only lead to a transportation catastrophe for this country as rail, waterway and public transport levels decline in real terms to the point where, when the great energy spree finally comes to an end, we are left in this country without any viable transport system whatever. And finally, the more roads and motorways we build, the more we commit this country to the desperate international struggle for increasingly scarce resources (including energy) to maintain the profligate and wasteful society that they so create and so exemplify – and that way is a short cut to world nuclear conflict.

It was because of all this that, early in 1973, I joined the National Transportation Working Party of the Conservation Society. I can hear its Chairman, Dr Leonard Taitz, allocating my territory now:

'You will take over the West Midland, then.'

Those words can be said to have changed my life. For only days afterwards a letter arrived informing me that in three months' time a public inquiry was to be held at Bromsgrove into proposals for a section of the M42 motorway. I cannot say that at the time I viewed the prospect with much enthusiasm, as it appeared to involve me in a great deal of work in a field of which I knew nothing at all. Before long, however, I began to discover that some very interesting matters were afoot. The Midland Motorways Action Committee (M.M.A.C.), an organisation linking many groups and societies throughout all the Midland counties, had, it appeared, under the leadership of Mrs Barbara Maude, Mr George Knott and others begun to assert that it was their right to object not merely to the *line of route* of a motorway, but also to the *need* for the motorway at all.

I well remember the tension and excitement of that opening day. It was a capacity attendance, and the conversation on all lips when I arrived was: 'Shall we be able to object to the need for the motorway?' I had thought it was all cut and dried, for I knew of a letter received by the local M.P. from the Parliamentary Under Secretary, Department of Environment, which included the unequivocal words:

> It will be open to objectors, if they so desire, to question the need for such a motorway, and there would, of course, be no question of excluding such arguments.

So what was the issue? The issue was (and this was as good an introduction to the world of highway planning as any) that an objector had subsequently received a letter, signed by the same P.U.S., including the words:

> There will, however, be no restriction on debating at the Inquiry the detailed basis of and need for the proposed new M42 route *on the present draft line*. (author's italics)

and a further letter, similarly signed, included the words:

> In all fairness, I must point out, however, that very extensive and detailed research into traffic flows has been carried out, and this indicates clearly that a new motorway route is needed.

There was not much point, under these circumstances, in expecting to be allowed to object on the grounds of need. And it was for this reason that the M.M.A.C. had instructed leading counsel to raise a 'matter of procedure' before the inspector, Major General Edge. I can see the scene now, with every actor on stage. Edge, handsome, personable, with the inevitable military air, banging his gavel: 'Good morning ladies and gentlemen . . .', Counsel for the Department, his junior, opposing counsel, the rows of official witnesses, documents piled, a projector, screen, the leading objectors . . .

A hunched figure rose, Mr Harold Marnham, Q.C.:

'Sir, my clients require me to raise a matter of procedure before you . . .'

He quoted from the 1959 Highways Act, Schemes Under Section Eleven, Section 10:

'After considering any objections to the proposed scheme that are not withdrawn . . . the Minister may make or confirm the scheme . . .' and from Section 6, which requires the Minister to give due consideration to the requirements of local and national planning. He concluded:

> With respect, Sir, there is nothing here in the relevant Act about policy that cannot be objected to . . . The words, Sir, are '*any* objections not withdrawn'. And if the Minister is required to take planning considerations into account, how then can need be excluded in an objection? It is my submission to you, Sir, that neither in equity nor in law has the Minister the power to deny grounds of objection to the public.

One could not but admire it. There was a brief silence. The inspector then turned to Counsel for the Department, Lionel Reed, who rose. His reply was brief and very much to the point, and included as I took them down the key words:

> An inspector is empowered to decide what evidence to hear or not. He has the power to disregard the expressed views of a Minister, but of course, he must act within the law . . .

There were further submissions, including a brief one from me, and the inspector then gave his judgement, which, clearly conceding the point, was as follows:

1. I will decide that which is relevant.

2. If Parliament has pronounced on certain matters, then these matters are not relevant.
3. I will not seek to be restrictive. I will not exclude any evidence to rebut evidence submitted by the Department. It is up to objectors to make their case relating to such questions as the petroleum shortage and their relevance to traffic counts, etc.

It was an inspiring experience. The might of government brought low, it seemed, by the citing of law and statute. Not being numerate, it seemed to me that here, in issues of procedure, was an area I might draw a bow in.

It was decided that the Conservation Society should present an objection. Our witnesses were three: a leading authority on energy, a leading agronomist and an economist. It should be noted that though this inquiry antedated the 1973 oil 'crisis' by several months, petroleum supplies were already a world concern. But during our presentation a surprising thing happened. Here was a leading authority on world oil giving evidence at an inquiry into a motorway proposal, the 'design year' for which was somewhere in the 1990s, and he had hardly started when Counsel for the Department packed up his papers and strolled out. And when at the end of the day I came to make my final submission, I could not but be struck by the bland indifference on the Department's side. And the Major General seemed hardly more interested.

It was then that it began to seem to me that the victory won for objectors by Marnham Q.C. was a hollow one indeed. The highways civil servants at Marsham Street had readily taken account of it. Let objectors spend their time and money on all the expertise they can muster: our inspectors will solemnly listen to it all; our counsel will ignore it all; and we, here in Marsham Street, will dispose of it all – in the waste paper basket. It was a jolly sort of arrangement keeping everyone happy – even the objectors, until their moment of truth a year or so later.

No, it seemed to me that the only interpretation of that legal victory at Bromsgrove was that *if objectors could challenge a motorway proposal on the grounds of need, then the Department must be required to propose the motorway on the grounds of need.* And, of course, if 'need' meant anything at all, it certainly meant more than a need to satisfy a traffic prediction. But this meant a parting of the ways. The M.M.A.C. decided to act strictly according to the letter of the law and to put their trust therein. I followed my own star and let it lead me where it would; and that my course has been a stormy one the following chapters will make clear. But in the end there has been a meeting, a synthesis. For in a curious way, as is made clear in the final chapter, attention once again centres upon the M42 inquiry; and the High Court action (*Bushel* v. *The Secretary of State for the Environment*) initiated by the M.M.A.C. plays a major part in the conclusions and findings which end this book.

1 The Long Haul

The public can never have the evidence and information upon which it can challenge a proposal.

A. G. Harcourt, one-time inspector into road proposals

The first thing that happened was the organisation of a walk-out. This took place on a day in September 1973 before Major General Edge at the inquiry into the M40 at Kenilworth. The cast was the same. I was supported by Mrs Maude and others representing organisations in the Midlands. It was my first procedural submission and effectively requested the inspector to adjourn the proceedings on the following grounds: The Department had always provided an objector who objected merely to the line-of-route of the motorway with evidence relating to the need for the motorway to take that particular line (and thus enabled him to formulate an objection). Therefore, now that objectors could object to the need for a motorway at all, it was incumbent upon the Department similarly to provide them with evidence justifying need (relating to general transport planning, energy, etc.), for without this, no objection could properly be formulated.

When the inspector refused, I recall slamming down the lid of my case and saying that, to show our contempt for these improper proceedings, we would walk out. Which we did. It was a very quiet affair, but significantly B.B.C. national T.V. had come at my request; we walked straight out upon running cameras, and there followed the first television interview. It could be said that battle had been joined.

The next issue was a month later and constituted a significant thread throughout the battle of the next three years. At Pamela

Johnson's suggestion, I intended to intervene in the M20 inquiry then taking place at Ashford in Kent with a submission (exceedingly weak) that, as the 1971 Town and Country Planning Act laid upon local authorities when proposing their structure plans the obligation to show adequate evidence relating to resources, then it was logical that the Department of the Environment should be required to give similar evidence when proposing motorways. The night before, it occurred to me that perhaps there was something to that effect in the

Highways Act. I was staying at Pamela's in Sussex and didn't have the Act with me, but I did have four photostated pages, and there then occurred one of those turns of good fortune which are so difficult to account for in ordinary terms. For out of one of them there leapt the following startling words:

Where the Minister (*a*) proposes to make a scheme under Section Eleven of this Act . . . the Minister . . . shall publish in at least one local newspaper circulating in the area . . . a notice (*a*) stating the general effect of the proposed scheme . . . (First Schedule, Section 11, Para. 7.)

'What does that mean to you?' I asked Pamela.

'I guess much the same as it does to you', she said.

Within moments, late though it was, we were before a barrister friend of Pamela's who lived nearby; she thought the idea was worth pursuing and, noting that the requirement was mandatory ('shall'), helped me with a useful precedent. Back home I burned some midnight oil with the result that the submission was ready for the following morning. In brief, it quoted Para. 68 of the House of Commons Expenditure Committee 'Minutes of Evidence' of 12 November 1973, where a Mr Pelling for the Department replied to a question as to how long it took to 'get a road going' with the significant words: 'The main problem is the protection of the rights of the individual through the statutory procedures. Ministers naturally must be concerned to protect those rights . . .'

It pointed out that the statutory procedures had not been complied with in the matter of Section 11, Para. 7 of the 1959 Highways Act. The published notice said nothing whatever about the general effect. The submission made a comparison with the requirements laid upon local authorities under the Town and Country Planning (Consolidating) Act 1971 regarding their Structure Plans (wherein the general effect of their proposals is clearly required to be publicly stated), and suggested that Parliament's intention was similar in respect of highway proposals.

It looked at the implications where mandatory requirements had not been complied with and concluded with a request that the inspector accordingly declare the inquiry null and void and that it be not reconvened until the 'protection of the rights of the individual through the statutory procedures' had been observed. (In extended form this submission is one of the seven submissions read before the inspector at the Aire Valley inquiry, and is in Appendix 2.)

What followed was interesting rather than dramatic. Counsel for the Department uttered the words which I was to hear so frequently in the ensuing years:

'Mr Tyme's remedy, Sir,' he said, 'can only lie with the High Court.'

And there followed a brief news item in *The Times* to the effect that the Conservation Society intended to take out a writ. But this represented our aspiration rather than our true position. Taking out writs, either of Mandamus or Certiorari was a vastly more difficult business than it seemed, and very expensive. However, that the issue was never proceeded with proved in the end, I think, advantageous.

* * *

It was not long after this that progress was made in the original direction indicated by the M42 inquiry. The Council on Tribunals is a body set up by Parliament in 1958:

(a) to keep under review the constitution and working of the tribunals specified in Schedule I to this Act . . .

and Para. 52 of the Do E booklet for objectors, *Public Inquiries into Road Proposals*, stated:

If you believe that the procedure followed was unsatisfactory, you can send details to the Secretary to the Council on Tribunals . . .

It seemed that here was my recourse, as I now faced two inquiries scheduled for the early months of 1974, namely inquiries into the M56 at Chester and the M65 at Burnley. Accordingly I wrote to the Secretary to the Council, and subsequently had an interview with him. The substance of my request, expressed in two letters, was much the same as that made before Major General Edge at Kenilworth. The reply was highly significant, and was to prove uniquely valuable in all of the ensuing struggle, and I take the opportunity now of acknowledging my debt to both the Council and its Secretary for its carefully worded contents. (The correspondence appears in Appendix 1.) The letter included this vital paragraph:

It may be of help to you to set out the Council's position in this matter. The Council endorse the view taken by the Franks Committee on Administrative Tribunals and Enquiries in 1957 that, whenever possible, some indication of the general policy relevant to the particular proposal under inquiry should be available before the inquiry. They consider that a witness or witnesses of the

However, the second of the inquiries referred to was of a great deal more significance, for at the M65 inquiry at Burnley events were set in train which were to lead directly to the three great battles which, in the space of one year, one annus mirabilis, utterly destroyed the Department's credibility. Two things distinguished it from the Chester inquiry. First, the case against the proposal was more obvious. Down that valley through which it was intended to smash a six-lane highway ran a railway and a canal, the former massively under-used, the latter not used at all. Secondly, a further string was added to the bow. Chester had raised in my mind the extent of the inspector's powers. Under what rules were the inquiries held to which one might have access? It turned out that there weren't any, and thus a third submission, readily appreciated by the public, was added (for who would even play a game of football without knowing the rules?)

When I arrived at the inquiry to view the usual intimidating scene, the first person I met was Harold Marnham Q.C., who was appearing on the opposite side for Lancashire County Council. Counsel for the Department was Mr Fay Q.C. (whose name appears more than once in this narrative). I had the previous evening begun the process (which later, when more time and care could be afforded to it, proved to be the key to success) of seeking to inform the local people, the objectors, of what precisely I sought to do on their behalf. I met no more than a handful, but one of them paid dividends, as will be seen.

I rose, 'To bring to your notice three matters of procedure, Sir', and proceeded to read the three submissions, which, in sum, began to look a respectable case. Mr Fay's response was predictable, except on the matter of rules. After turning a lot of papers up, he had to admit that there weren't any. The inspector then made the interesting announcement: 'I will consider whether to adjourn the inquiry over lunch.' He invited me and the two opposing counsel to re-state our cases in his private room. At 2 p.m. the inquiry was reconvened and he pronounced the formula that was to become so drearily familiar to me over the years to come: '*Mr Tyme, I will report all these matters to the Secretary of State; meanwhile the inquiry will proceed.*'

There then occurred an electifying moment. Suddenly a man, one of the small group I had met the previous evening, strode angrily to the witness box. 'You'll do nothing of the sort!' he shouted. 'This inquiry can't proceed without rules. I demand them as my democratic right!' Expectation hung in the air! But there was no support of any kind from the hundreds present. In the silence the inspector gathered his wits. Mollifying words were uttered and after some further angry argument the man withdrew. The moment had passed. That afternoon, inch-high headlines stared from the front page of the local paper: 'INQUIRY STORM OVER RULES'

But the real significance of these events was little noted and had nothing to do with headlines. At the end of the day two gentlemen approached me. 'We are from Bingley in the Aire Valley,' they said. 'Perhaps we can have a talk . . .'

* * *

But I was not yet done with the M65 inquiry, and that all-important thread leading to the Aire Valley will be taken up later. I wished particularly to test once more, as I had later done at Chester, the presentation of an *objection* within the terms of the Council on Tribunals' interpretation. Accordingly, I presented an objection with two witnesses, one on energy and the other, Dr John Adams on traffic prediction methods. It was he who first and most devastatingly drew attention to the absurdities of the Department/road lobby's most valuable card, their statistically mystifying method of predicting future road traffic. It is important to note that *his attack was first made at Burnley; but there it devastated nothing, because that inquiry, unlike the M16 to follow, drew no national attention.* But back to matters of procedure. What happened when I presented this objection is best described in extracts from an account I sent, describing the events, to the Secretary to the Council on Tribunals:

Counsel for DoE I think Mr Tyme is under a misapprehension. In the letter from the Council on Tribunals to which he refers, we find the term 'the policy applicable'. Well, Sir, the policy applicable was the subject of my opening address. The Secretary of State's intention to build this highway . . .

Conservation Society I beg to disagree. If Learned Counsel will look at the preceding paragraph of the Council on Tribunal's letter, he will see the words 'general policy'. This clearly indicates the overall transport policy within which this motorway proposal is presumably set. Furthermore, as I recall, Learned Counsel gave indication that he was in dispute with me on the interpretation of this phrase on the opening day of this inquiry and I respectfully submit that the only way properly to interpret the disputed phrase is from the letters I wrote to the Council to which their letter is in reply. I have prepared copies and am willing that they should be read –

Inspector I think you should put your questions, Mr Tyme.

Conservation Society Well, Sir, my first question concerns the future of British Rail. In order adequately to understand this proposal we need to know whether the last 20 years of British Rail experience in terms of closure and contraction is going to continue, or whether, as was suggested by Mr Peyton in the House of Commons last December, the policy is going to be reversed – in which latter case, of course, the whole future of movement in the Calder Valley will look substantially different from that seemingly envisaged in the proposals. What is the latest position as the Department knows it?

Counsel for DoE I certainly do not intend to put up a witness to answer questions about railways.

A silence.

Conservation Society Well, Sir, it seems that I must accept that the Department is unwilling to explain how this proposal fits in with general transportation policy. I accept this position with regret, and give notice that I shall be referring the matter to the Council.

My regret was not too deeply felt. Any clearer indication of the Department's obdurate refusal to comply with the clear requirement as set down in the Council on Tribunals' letter could hardly be wished for.

* * *

The M16: the Motorway through Epping Forest

The M16 inquiry at Epping was characterised throughout by betrayal. The simple equation:

$$\text{Truth} + \text{Courage} + \text{Numbers} = \text{Invincibility}$$

was not enunciated till later, but the principle therein was understood. *If any of the three left-hand factors were absent, then all would fail.* At Epping the absence of the third factor ensured that the battle of the M16 was effectively lost. Insufficient numbers of people understood what it was all about, and what people cannot understand they can hardly be expected to take risks for. What happened was that in the early summer of 1974 I was asked to explain my proposals to a meeting of the 'Alliance Against the M16', an umbrella organisation incorporating a very large number of societies and action committees covering the whole area. I think, looking back at that meeting, that my first impression is unaltered, and that the information was well received by the thirty or forty representatives present. I was assured by the steering committee that further meetings of the various memberships would be organised, so that everybody would fully understand the issues before committing themselves to action.

Those meetings were never held. Time and again, as the summer and autumn drifted by, I rang the committee member responsible. But to no effect. Meanwhile the committee went ahead with the traditional approach and briefed counsel and gave every indication of their intention to play according to the rules from the beginning. (I say from the beginning, because this line could always be later resorted to should the procedural attack fail.) It is useful at this point to explain the significance of this particular inquiry. The six-lane motorway was planned to slice through one of the last remaining segments of Epping Forest, one of the few great 'relict' forests of England, granted by Queen Victoria to the people of London for perpetuity – a unique national asset. The national concern was there; the whole of the media was sympathetic. It was an unprecedented opportunity. It may be similarly useful at this point to ask: but what, by the proposals you made, did you hope to achieve? What was the object and the measure of success? The answer is that already it was clear that changes could only be brought about by a change in the political will as expressed by Parliament. In the face of the (corrupt) alliance between the road lobby and the highway mandarins in Marsham Street, civil disobedience was, I had become convinced, the only means of showing the extent and depth of popular feeling and opposition to the endless proliferation of motorways and roads and the destruction of town and valley, woodland and farmland and the ruin of this nation's future which was its inevitable concomitant (see the forthcoming *Roads to Ruin* by John Tyme to be published by Macmillan).

Cut off from the people of the area, I had to resort to obtaining representation from as many organisations across the country as I could muster and by the opening of the inquiry I was acting for no less than nine organisations from Northumberland to Devon.

Meanwhile, however, significant meetings were being held a very long way from Epping Forest. For the contact made at Burnley had borne fruit, and I was asked to meet the Committee of the Aire Valley Protection Society. Here I met a group of people who had no inclination whatever to play the game according to the rules. 'Well,' I said, 'send representatives to the M16 inquiry. You will thus do two things: give help there where it is likely to be needed, and find out what it is all about for yourselves.' It was agreed that a group would travel down.

On 3 December 1974 at a hall in Epping Village at 10.30 a.m. the inspector, a Mr Clinch, opened the inquiry. The events of the two-day episode are as follows. After his opening statement, and at the point where he asked Mr Newey, Counsel for the Department to commence 'appearances' I rose, requesting to raise 'certain matters of procedure'. This granted, I read the three submissions. Counsel for the Department then rebutted them, and I answered the rebuttal. The inspector then (inevitably) announced: 'While I shall report the matters to the Secretary of State, the Inquiry will . . .'. He got no further. Uproar erupted from the 300 or so present. Do E officials were, I was told, chased out of the room, and an impromptu meeting was then organised at which I was asked to speak. A resolution was passed nem.con. to the effect that the inquiry should not open until the three matters were resolved. As the local *Gazette* of 6 December had it: 'Immediately after the inquiry broke up for the last time, counsel and solicitors on both sides adjourned to a small room at the back of the hall to discuss what was described as an "unprecedented situation".'

The unprecedented situation was something of a success, but we were now to reap the fruits of betrayal. At 11 a.m. on the following day, the inquiry reconvened. There were barely fifty people present. Most had gone back to work and the Aire Valley people had returned north. *The equation was unbalanced.* The inspector opened with a statement to the effect that all three matters could only be remedied in the High Court (this was true only in respect of the matter of the Statutory Notice; his discretionary powers could well incorporate the other two). This was followed by such honeyed words of reassurance that you might have thought him a lifelong friend of every objector present. Lack of understanding – and thus conviction – on behalf of the people present ensured that this balm was inadequately understood. Then, quite suddenly, counsel for the Alliance rose. 'Sir,' he said, 'I represent a major objector to this proposal. I had the benefit of reading Mr Tyme's submissions some time ago, and I confess I didn't think much of them. My clients wish to put their case, and the inquiry to continue.' From that it was but moments before Mr Newey, Counsel for the Department, rose for his 'appearance'. The inquiry was under way.

I was consumed with almost hopeless anger, and I determined to take upon myself some gesture of civil disobedience. But how to do it?

I then remembered a story of a Cambridge undergraduate who, on being compelled to read a lesson in chapel, had reluctantly done so but instead of stopping at the appropriate verse had completed the chapter, read the next and got well into the next before being forcibly removed. So, when the inspector returned after lunch, I said to him that as he clearly had not understood my submissions, I intended to read them again. Despite his angry refusal, the ensuing uproar, an adjournment and the summoning of ushers, I continued reading until the police arrived to escort me out. *This was the first time the police had been summoned to a motorway inquiry. I was determined it should not be the last.*

Nevertheless, over a cup of coffee the following morning at St Pancras Station I confess to breaking down. Failure seemed complete. It was not so, however. The New Year was to see significant developments. On 10 January no less eminent a counsel than Mr George Dobry Q.C. appeared for the Conservation Society at the Epping inquiry on the matter of the 'general effect' notice. As reported in the Civic Trust *Newsletter* for May:

> The Conservation Society was advised by Mr George Dobry Q.C. that the notices required under the provision of the Highways Act 1959, and published by the Secretary of State in a local newspaper, announcing the proposal to make an order for a road scheme 'were seriously defective and do not constitute a performance by the Secretary of State of his duty under the First Schedule, Paragraph 7 of the Act'.

This led to renewed national interest and a major centre-page article in the *Financial Times*, which was the first major article hostile to the Department to appear in the responsible national press. The occasion was recorded in a *Guardian* article 'Motor Fray' by Robert Waterhouse, from which the following is taken:

> Outside there was a half-hearted attempt at a public meeting where Mr Tyme produced a contingency telegram addressed to the Prime Minister, requesting Mr Wilson's 'personal intervention to end this total contempt for law by a department of state which no longer acts in the national interest'.
>
> Supporters managed an impromptu whip-round for the cost of the telegram and Mr Tyme was last seen disappearing, duffle bag over shoulder, in the direction of Epping Post Office.
>
> Such are the almost inevitable elements of pathos when lowly individuals challenge the great machinery of government . . .
>
> Mr Tyme pursues his goal like a latter-day evangelist, secure in his doctrine

of environmental sanity. But it may take a miracle of a kind he hasn't yet produced – like really stopping the next motorway inquiry – to convince others that those atheists in government have deigned to listen.

The 'miracle of a kind not yet produced' was not far away. It is necessary to pause here, however, to describe significant new ammunition which was coming to hand. *I had noticed that Mr Clinch at Epping, taking his cue from Counsel for the Department, had been at great pains to explain that the inquiry was 'not the proper place for Mr Tyme to raise these matters'. The Council on Tribunals' submission by inference brought into question government policy. This, Mr Clinch pointed out, had been settled by Parliament, and it was 'to that body that Mr Tyme should have recourse'. I decided to find out to what extent this statement was true.*

This was to require a great deal more reading than I am normally accustomed to (in the end 146 thick bound volumes of Hansard) which was undertaken in the Sheffield Central Library. By the end of it I had every reference to transport, motorways, truck roads and highways ever made in the House of Commons over a period of ten years. The interesting results and relevant abstracts are attached to the Winchester submission in Appendix 3.

During the early summer months of that year (1975) I attended two inquiries within the space of eight days, one at Byfleet in Surrey (into a section of the M25 Outer London Orbital) and the other at Llandudno (the A55 proposals, an interesting exercise in deciding which way to wreck three Welsh seaside towns). At neither was there any local support, and although I was now acting for no less than thirty-five organisations across the country, my attendance had no more intent than to 'put the Department on notice'. It certainly did no more, though one thing it did was to raise certain questions about my ability to attend so many inquiries while employed as a full-time lecturer at Sheffield Polytechnic. Thus was a shadow cast before.

These two inquiries might be said to mark the end of the preliminary stage of the long struggle. Attention now turned to the Aire Valley Trunk Road proposal, and this issue requires a chapter of its own.

2
The Battle of the Aire Valley

We shall this day light such a candle by God's grace in England as, I trust, shall never be put out.

Bishop Latimer

Before commencing this narrative, it might be appropriate to set down a number of principles by which I have sought to guide my conduct. *The first* was well expressed by Lord Milner in 1909, when he said:

If we believe a thing to be bad, and if we have a right to prevent it, it is our duty to try to prevent it and to damn the consequences.

The second I shall have to put into my own words:

It is not merely the right but the duty of every citizen to commit a minor crime in order to prevent the commission of a major.

The opening of a road or motorway inquiry under the conditions currently prevailing is a very major crime indeed. That is my firm belief and I act upon it. Of course, applying this principle does not protect one from the consequences of one's actions. Knowing that a gang of thugs is intent upon a mugging requires one to take appropriate measures and, all other measures having failed, to attack before they do. And that may very well incur a charge of assault and battery, which could be very difficult to defend. And *the third* principle is that from my reading of history I see our democratic way of life (which, for all its faults still makes this country the refuge for innumerable victims of persecution and arbitrary government) not as something that was handed to us on a plate, but as something fought and suffered for over centuries, and which, therefore, deserves constant vigilance in its defence. *I have, therefore, no conscience about noise at an inquiry or organised sit-downs. I see them as very small beer when contrasted with two things – the bitter nature of the historical*

struggle referred to, and the very real threat to many of the freedoms and privileges that we have for so long taken for granted coming from massive bureaucratic and technocratic powers which can so easily undermine the democratic decision-making process.

* * *

The organisation of the Aire Valley Protection Society (A.V.P.S.) was highly efficient. The number of objectors ultimately was well into four figures. Feeling ran high. I confess to experiencing great anxiety at the time, fearing that the Department, putting two and two together – the efficiency and strength of the movement *and* the possibility of a major procedural battle – might substantially defer bringing the proposal to inquiry. I therefore sought publicly to give the impression that, disillusioned by the last two inquiries, I had given up. In pursuance of this I did not attend a very important inquiry in Manchester into the M67 Hyde by-pass under Rear Admiral Nixon. This nail-biting period extended into the autumn and I can remember numerous telephone calls to farmer John Burnhope always seeking news of the announcement. In due course, however, we were able to relax; the inquiry date was fixed for 4 November. Relax is perhaps not quite the right word. Release from anxiety certainly, but relaxation it was not. We were galvanised into feverish activity. The lines required to be drawn for what was clearly to be the decisive battle.

The meetings had commenced in the late summer and from about the end of August my average sleep of a night was between three and four hours. There seemed to be something about the hour of 4 a.m. In all the days of September and October it was rarely that I did not wake at that hour with some clear vision of what should be done and how.

The valley was united. Separate meetings were held in different areas and the issues and tactics clearly explained. The committee and leaders of the A.V.P.S.: Graham Carey, John Burnhope, Harold Sutcliffe, Ron Craven, Joan Miller and others were as solid as a rock. As the meetings took place (*with everyone guaranteeing the probity of everyone else*) problems were ironed out. Speech was free; doubts were many, and each had to be resolved. No two people see a matter in exactly the same light. An infinitude of patience was needed, together with equal tact as, iron-hard, *the clear and grim intention was spelled out: We would, by all and every means short of violence, prevent the inquiry from opening until all matters had been resolved.* The possible role of the police was discussed; legal aspects were gone into; the vexed question of maintaining numbers over what could turn out to be a long struggle (to maintain the vital equation) was tackled, and people undertook to prepare rotas. The need to ensure maximum media publicity; the need to assist people to verbalise their own support; the possibility of opposition from the floor drummed up by the local road lobby; the danger of *agent provocateur* inspired violence: all these and other matters were openly dicussed. The submissions once in their completed form (which Pamela Johnson agreed to read should anything happen to me) were read out and explained to all. *None went away unhappy; morale was high. Immutable determination was the keynote. It was an exhausting but inspiring period.*

And so the weeks passed and the day approached. One of the 4 a.m. ideas was that I should obtain signatures to a letter authorising me to act on behalf of the signatories on the specific issues set down therein. This had two objectives. All who signed, in that they were required to read the contents very carefully, knew precisely what the issues were. Secondly it was designed to delineate very clearly my authorisation. This proved to have been a wise precaution when the Treasury Solicitor's representative asked for copies. In the end 183 objectors signed the letter, the main substance of which is set down below.

Dear Mr Tyme or Mrs Johnson

We, the undersigned Objectors, write to ask you to represent us and to act on our behalf on the scheduled opening day (and any necessary day thereafter following) of the Public Inquiry into the Aire Valley Trunk Road proposal.

We require you, in so acting, categorically to demand of the Inspector appointed to hold the Inquiry that the said Inquiry be NOT OPENED until the Department of the Environment and/or the North Eastern Road Construction Unit acting on its behalf have complied with the Law and accepted constitutional practice, and thereby have:

1. Complied with Paragraph 1 of Part I of the First Schedule, . . . etc.
2. Published a set of Rules of Procedure, . . . etc.
3. Complied with the expressed views of the Council on Tribunals, . . . etc.
4. Obtained full Parliamentary approval, . . . etc., etc.

In particular, we request you to ensure that these four matters are raised PRIOR TO THE TAKING OF APPEARANCES OR ANY OTHER FORMAL PROCEEDING; this for the reason that precedent shows that, when these issues have been raised AFTER such formalities, inspectors have seen fit to regard the inquiries as having been officially opened. We therefore reiterate: this Inquiry should not open with any proceeding whatsoever until these matters have been properly resolved. This letter authorises you so to act on our behalf.

* * *

At 10.30 a.m. on Tuesday 4 November 1975 the Inspector, Mr Ernest Ridge, sat down at his desk on the platform of the assembly hall of Shipley School and bid the packed hall good morning. He was greeted by silence. Then, when he started to introduce himself prior to reading his authorisation by the Secretary of State, there occurred a deafening shout: 'NO!' and there followed a quarter of an hour of points of order, shouts and speeches demanding that I be heard. Amongst it all the following exchange took place:

Inspector My name . . .
Uproar.
Inspector But I only want to tell you my name.
John Burnhope Right then, who are you?
Inspector My name is Ernest Ridge and I am . . .
Uproar.
John Burnhope Mr Ridge, you have told us your name. That is what you said you wanted to tell us and now you have told it us. We don't want to hear any more. We now wish *you* to hear Mr Tyme . . .
Cheers and applause.

The Inspector then adjourned and on his return announced that he would hear me. The time was 11 a.m. precisely.

The submission* (see Appendix 2) took an hour and a half to read, it being in seven parts, much of which consisted of considerable detail regarding the situation following the Epping inquiry. *First* I submitted that the Secretary of State had failed to comply with his obligation in the matter of the Statutory Notice. *Second* that there were no Rules of Procedure, from which it followed that objectors were first denied protection and second denied their rights in law. *Third* that, in failing to comply with the Council on Tribunals' interpretation of its duties, the Department had denied objectors their only means of understanding the proposal and had thereby

* A detailed reading of the submissions is essential to any full understanding of the position adopted and its rigorous defence throughout the inquiry.

failed to comply with a ruling given by the only authority to which recourse could be had in matters of dispute regarding inquiry procedure. *Fourth* that the Secretary of State had, in his total neglect of other modes of transport, failed to comply with the 1959 Highways Act in the matter of national planning. *Fifth* that the Secretary of State had failed to comply with Section 20 of the Act in the matter of navigation over the waters (the Aire and Calder Canal runs along the Aire Valley). *Sixth* that the Secretary of State had failed to ensure adequate Parliamentary control of the road programme. And *seventh* that the road programme represented a corruption of the function of government by the Secretary of State's Department. (This all-important matter is dealt with at length in Chapter 5.)

Immediately following the submission, I requested the inspector to report directly to the Secretary of State on two matters. The more significant of the two was read out as follows:

> On 5 May 1972, several of my clients, members of a society in Keighley, visited the Road Construction Unit at Harrogate. They took their Society Minute Book with them. One of their members had suffered a break-down and had spent a period in hospital because of the threat to her home posed by the road. The engineer who interviewed them used the following words: 'You people are inanimate objects. We will fill the hospitals if necessary. What you people have got to understand is that the road is going through.'
>
> Those words were set down in the Minute Book immediately afterwards and my clients will swear affidavits regarding the accuracy of this account. I request you, Sir, to require from the Secretary of State an immediate investigation of this matter.

It proved impossible, in all the complexity and confusion of the ensuing events to follow this matter through with the rigour that it deserved.

Counsel for the Department's rebuttal followed, but by the nature of the circumstances, it could not be in any way effective, for in saying that the matters raised were beyond the scope of the inquiry and entirely outwith the Inspector's powers, he could not but play into our hands. My counter-rebuttal was followed by the Inspector's decision that he would report matters to the Secretary of State, but that the inquiry would not be adjourned as I had requested.

This unleashed pandemonium. During a short adjournment the police were called for the first time, to deny Mrs Pamela Johnson a hearing. There was uproar:

'On what authority, Mr Ridge, have you called the police to this meeting?'

'By what law?'

'By what Rule?'

'Answer! Answer!'

A further adjournment followed during which the police were helpfully invited to arrest the inspector and any members of the Department of the Environment they could find for 'conspiracy to contravene the 1959 Highways Act.' The police declined to do so. A senior police officer arrived and was invited onto the platform, but refused the invitation and shortly afterwards withdrew his men. Early in the afternoon the inquiry was adjourned to the following day.

The second day was no different. During a lengthy adjournment a public meeting was held within the hall and a chairman elected. Speeches were made and a resolution proposed by Mr William Moore: 'This meeting declares that this public inquiry shall not open until all matters raised by Mr Tyme yesterday have been settled.' According to the *Guardian* report of 6 November: 'About 120 objectors voted for the resolution and only about three against.' Further attempts to open the proceedings met with more points of order, interventions, singing and cheers, with demands for the police, who had reappeared, to remove Mr Ridge from the hall because he 'cannot be independent when he is paid by the Secretary of State for the Environment'. A further adjournment was followed by the same, including three cheers for the police 'for behaving perfectly properly'.

At the end of the day a meeting was held in a nearby church hall. There, as the *Guardian* reported, I reminded the objectors of John Hampden who stood against the tyranny of Charles I and said: 'We have in this room 120 John Hampdens. Make sure on Friday there are 240.'

* * *

The Aire valley 'inquiry' lasted for a period of ninedays, spread over three separate periods from November 1975 to the following February. The adjournment after the early days of November marks the end of what might be called the easy-going stage. The second phase, commencing later in November, was characterised by heavy police action, arrests and summonses. Throughout there were nine unscheduled adjournments. Two police forces were involved and at one time it was estimated that eighty members of the West Yorkshire force were present. There were twelve arrests and many summonses which resulted in the end in heavy fines totalling several hundred pounds. A sinister development was the use, in the final phase, of persons from a security firm as stewards. A document which came into our hands authorised them as follows:

> If the person or persons refuse to leave, then stewards should attempt peaceably to remove them. Sufficient force should be used to indicate an intention to remove but not sufficient to hurt.

For us the implications of this were serious in the extreme. Under the circumstances the fact that no violence was done to any person by any objector, despite the extreme provocation to which so many of them were subjected, is a remarkable testament to their self-control. On one occasion when the inspector in an attempt to force the opening of the inquiry read over his opening statement with the loudspeakers turned up to a deafening level (which was added to by the counter-noise of shouting and banging chairs) some people took matters into their own hands. When Counsel for the Department took over the microphone and began his statement, a woman sought to seize it and was herself seized. There were no police present and it was a moment of maximum danger. I was fully engaged in seeking to restrain people when I suddenly found a face thrust close to mine and heard the words: 'You are responsible for all this. I am going to get you!' When followed by others, this person was seen to enter a door marked 'Private: Official Personnel Only' and it was discovered subsequently that he was a police officer, at the time off duty. An investigation by officers of another force was later held. On another occasion, this time outside the inquiry, I was warned by a senior road lobby official that I would be put in a 'Peter Hain situation', which he kindly defined for me: '*It means framed. That is what happens to people like you who cause trouble for the Establishment.*'* We were all under the greatest strain. Many feared for their jobs. None of us could know where it would all end.

It was certainly one of the grimmest periods of my life. I remember the stark foreboding that was in those cold grey dawns at John

* Within hours of this sobering, not to say frightening, episode I had a sworn statement in the records of the South Yorkshire Police Headquarters, Sheffield, and sworn affidavits lodged in my bank and elsewhere.

Burnhope's farm. I remember in the later phase a meeting of objectors on the eve of the re-opening when it had first become known that security firm men were to be present. I recall the anxiety that people knew, and I remember that my final words to them were taken from *Richard III*:

Go, Gentlemen, every man unto his charge.
Let not our babbling dreams afright our souls;
Conscience is a word that cowards use to keep the strong in awe.
Our strong hearts be our consciences, words our law!
March on, join bravely, let's to it pell mell,
If not to heaven, then hand in hand to hell!

(I had substituted 'hearts' for 'arms' and 'words' for 'swords'.)

I remember vividly on the following day the inspector addressing those same security officials: 'Carry Mr Tyme out!' and my being surrounded at once by half a dozen stalwart Yorkshiremen: 'Do not touch him!' and I was left alone.

I had to accept moral responsibility for all that happened, but no more than anyone else could I know where the day would lead us. But it was no use flinching. *We had set ourselves to oppose the concentrated power of a major Department of State, and must take what came to us.* Courage and bright spirits were everywhere. The game was afoot, we would play it out to the end.

The end came dramatically enough on 4 February. It had been public knowledge for some time that, should trouble ensue at the

February re-opening, the inquiry would be continued behind closed doors, with the proceedings relayed by loudspeaker elsewhere to objectors. The plan (known as Plan B) was to continue the inquiry across the road in Shipley Town Hall Council Chamber. At 11 a.m. on 4 February, the inspector, tired of what he called 'a barrage of noise, abuse and other distractions', announced his decision to invoke Plan B. He adjourned until 2 p. m. in the Council Chamber.

Precisely what happened thereafter I am uniquely able to recount. I had, the previous day, reconnoitred the Town Hall, a converted Victorian house with a small entrance and vestibule, a twisting stairway and a narrow upper corridor at both ends of which were doors leading into the chamber. I was convinced that plain common sense in the interest of public safety would ensure precautions to prevent large numbers of people entering the building. Accordingly a major demonstration was planned to take place in the street with objectors demanding access to their 'public' inquiry. At 2 p.m. when the voices started coming over the loudspeaker, I made no effort to leave the hall in advance of the very large and angry crowd. I was, in fact, one of the last to leave. To my great shock, when I reached the street, it was to find the last of the objectors entering the Town Hall door. I raced over, but when I reached the upper corridor, I could for the moment get no further. A steward, reputed to have broken ribs, but in fact only bruised, was being carried out. The effect, of course, was like forcing water down a narrow pipe, closed at the end. The doors to the Council Chamber were not locked but, incredibly enough, jammed on the inside with chairs under the handles. As a result, they gave ever so slightly on pressure. And the pressure increased as more and more people poured along the corridor.

The chairs splintered, the handles snapped, and the first person to come flying into the Chamber was Mr John Burnhope, pig farmer. Almost immediately the room filled with objectors who began singing 'We shall not be moved'. When I finally gained access, it was with the intention of defusing the situation at the earliest. 'We have made out point,' I said. 'We shall now remove ourselves.'

* * *

It was the end. Quite apart from events described, things had been happening behind the scenes. The day before, when numbers entering the inquiry had been restricted by the security personnel, three gentlemen had arrived well before 10 a.m. Knocking on the door, one of them said to the steward: 'We are three councillors of the City of

Bradford and I am the Deputy Lord Mayor. We require to attend this inquiry.' As reported to me by a witness, the steward replied with the immortal words: 'I don't care if you are Harold Wilson. You are not coming in here.' Urgent telephone calls to Whitehall were said to have followed. Press reportage had been almost unanimously favourable. No one likes unruly behaviour and this had been fully reported, *but the reasons behind it all had been and were to be fully and fairly explained.* A *Guardian* leader for 10 February, headed 'Road to Pandemonium', included the words:

The Yorkshire objectors have won the battle, but are unlikely to win the war with such tactics. There are indications however that government too may be recognising that the scales are weighted too heavily against the public. Such recognition, to be helpful, must lead to public and Parliament having a fuller and earlier share in the road planning procedures, with the national context of any proposal being made apparent before inquiries into individual schemes begin. Already the Council on Tribunals has been asked to review inquiry procedures. What the Secretary of State should also be considering is two-tier participation, with discussion of need and context of a future motorway preceding any debate on more specific details about its line. An arrangement for meeting some of the costs serious objectors incur must also be devised. If such ideas are pursued then at least the people most affected will no longer have one hand tied behind their back in the contest against bureaucracy; and progress will have emerged from the shambles at Shipley.

A centre-page article in *The Times* for 14 February by Michael Horsnell entitled 'How the Aire Valley rough-house struck a blow for democracy' gave, as its title indicates, a full account of the implications of the long struggle. Television reportage (always worrying: they require visible action but are never able to give interview time adequate to the complexity of the case) had been wide and fair, and the B.B.C. had produced a major *Man Alive* film, 'The Battle of the Aire Valley', with dramatic film coverage followed by studio interviews and discussion involving the Minister and many interested parties.

It is interesting, furthermore, that for all the public condemnation of all those courageous people involved as hooligans and the like, people were not slow to take advantage of the new situation. For example, the Headquarters of the Council for the Protection of Rural England sent a letter to the Department of the Environment's North-East Road Construction Unit as follows:

I am advised by our West Riding County Branch that it would be most unwise to dismiss the unruly objectors who have disrupted this Inquiry as a bunch of cranks or to ignore the genuine and understandable feeling behind their tactics.

We suspect that the Inquiry is likely on its resumption to continue to encounter difficulties unless it is reconstituted on a new basis..

Many objectors at the Inquiry (our County Branch included) believe that no decision resulting from it can be seen to be reasonable unless the Inquiry's remit is widened to enable the Aire Valley Scheme to be considered as part of the much larger transport picture. To be more specific: we consider that the Inquiry should not be held until such time as a thorough comprehensive regional transport study is available and unless the associated Kirkhamgate–Dishforth and Shipley–Leeds roads are considered at the same time.

The postponement of the resumption of the Inquiry presents the opportunity for this to be done.

Christopher Hall (signed)
Director

The announcement of the abandonment of the inquiry was made on 6 February.

But accounts had to be paid. The final chapter came in late February 1976 and was the court appearances at Bingley Magistrates' Court. Many were heavily fined. Together with many others, however, I was merely bound over for a period of one month for the sum of ten pounds. I knew then that it was all over. On the return railway journey to Sheffield I suddenly began to feel ill. In a sense this was not surprising. During the long adjournments I had attended on the invitation of local objectors two inquiries as far apart as Penzance and Colchester, and two inquiries – in Lancashire and Sussex – where I represented the Conservation Society. These had been a great strain, added to which, in taking police advice regarding the threats that had been made to me, I had not lived at home for a long time, and for a period of six weeks had not slept in the same bed for more than four nights running. Influenza followed and I was for a while in a rather depressed and exhausted state.

3
The Battle of Winchester

So we question not: but every true English man that loves the peace and freedome of England will concurre with us.

An Agreement of the People, 1647

With the advancing spring, health and spirits returned in time for the first of the two major battles which were to dominate the summer and autumn of 1976.

To understand the first of them we must go back to the M42 inquiry at Bromsgrove. It will be remembered that there it had been established that it was an objector's right to object to a motorway on the grounds of need. There are basically two types of road/motorway inquiry. The first is a 'Line Order Inquiry' which does just what it says, inquires into the proposed road along its proposed line or route. Then, once the Line Order is signed (or made) by the Secretary of State, there follows a second inquiry. This is a tidying up operation. The road has been approved to smash its way through town or countryside, but there remain problems of the compulsory purchase of property in its way, the stopping up of roads and of course the infliction of further destruction in the form of slip roads, spurs, split-level junctions and so on. All these matters are subject to public inquiry under the provisions of the same Acts, the 1959/71 Highways Acts and the Compulsory Purchase Acts. They are called 'Side Road Order Inquiries' and/or Compulsory Purchase Order Inquiries.

The inhabitants of Winchester and its environs had, in 1971, experienced a Line Order Inquiry into a section of the M3 which would, if built, amongst other things, ensure ten lanes of highway across the Winchester watermeadows, and, by the traffic it would generate, do for the city of Winchester what roads have done for the once beautiful cities of Chichester, Salisbury, Gloucester, Exeter,

Worcester *et al.* Despite prodigious effort, the objectors had, by obeying the rules, merely found in 1975 that the Line Order was made. The Side Roads and Compulsory Purchase Order Inquiry was scheduled under Major General Edge for 29 June 1976.

In the intervening years a very powerful umbrella organisation, the M3 Joint Action Group, had been formed and it was faced with the problem of what to do. Two small events, both unknown to me at the time, tended to help it along. At an open meeting early in the year Professor Terence Morris had uttered the following nine words: 'Nothing,' he said, 'has been the same since the Aire Valley.' Secondly Mrs Denise Loveday, who was to be my hostess during the inquiry, read in her husband's copy of *New Civil Engineer* an account of the Aire Valley battle. That is the man – alluding to me – she said, we must get in touch with. Possibly because of these two events, one morning, in April I think, I received a letter inviting me to meet the M3 Action Group Committee. It was my first visit to that city of ancient kings since boyhood. The meeting was held in one of the great houses of the College. I told them very simply the following:

1. If you follow the rules you will go down again.
2. What the Aire Valley folk showed is that given fortitude and determination you can win.
3. Don't imagine it is all noise and clamour; it is an intellectual game of some sophistication.
4. You were, as I understand it, denied your rights under the Act in 1971.
5. I will work out the issues on which a stand can be made.
6. You must ensure maximum understanding of it by as many people as possible (I mentioned the equation).

Shortly afterwards I received a letter agreeing to a procedural battle. The stage was set. But there was a prelude, a prelude having all the hallmarks of tragedy and pathos. With issues in both cases precisely the same, the battle of Hornchurch was lost while that of Winchester was won. Why this was so, the following short narrative explains.

* * *

Defeat at Hornchurch

A month before the Winchester inquiry a similar inquiry was scheduled into the Havering section of the M25 at Hornchurch. Here the objectors lacked one crucial advantage common to both Winchester and the Aire Valley – a homogeneous community. The threatened area (London Green Belt land) adjoined a part of the amorphous outer London suburbia. However, the unremitting efforts of Lesley Lovelock, the Miltons, the Nelsons, the Popes and others had gathered about forty people to a public meeting which I addressed.

An important point must be made here. The people of Winchester were denied their rights under the Act to object on the grounds of need as far back as 1971. But the Havering Line Order Inquiry took place in 1973 *concurrently with the M42 Bromsgrove inquiry*. In other words, the Winchester people had been denied a right they hardly were aware of. The Havering objectors, reading their newspapers about the Bromsgrove ruling, had sought the same right for themselves *and been denied it* by Counsel for the Department, Mr Fay Q.C. and the Inspector, Mr Rolph. Their case, in other words was stronger than that of the Winchester objectors.

As a result of the public meeting forty or more people attended the inquiry. I read a lengthy submission, almost identical to that at Winchester, and as I had no wish for its contents to be accurately noted by the Department or the Treasury Solicitor I refused a copy to either of them or to the inspector. When the inspector refused to adjourn so that need could be considered, all forty or so people exploded in a manner which disposed of any doubts Aire Valley people might have had regarding the 'effete' southerners. But there were only forty of them, and over a two-day struggle they were all ejected by the police, and the inquiry, now empty of the public, continued. It was the unbalanced equation again. Nevertheless, it made considerable impact in the media, and led to an editorial in the *Evening Standard* which began with the words: 'The Tyme machine has arrived in outer-London.' *If there is such a thing as justice, then the wrong done to these people requires to be righted.* Fortunately, by the vigilance of one of the objectors, Miss Lesley Lovelock, who discovered that the Department had not complied with the requirements of the 1938 Green Belt Act, a further inquiry requires to be held, and thus a chance of putting matters to rights is to occur.

Before leaving this event two significant things require mention. First, like the Aire Valley people at Epping, so Winchester sent two important observers to Hornchurch in the persons of Mr Jock Macdonald, a College house-master, and Professor Terence Morris. Their experience proved invaluable. And secondly the reply to my submission by Counsel for the Department included these significant words:

Whatever Mr Tyme says can be of no significance. The Secretary of State has signed the Line Order. It is made. It is a fact of law. It cannot be unmade and the issue cannot be re-opened.

How true this was to remain, when the balance of forces shifted and the vital equation came into play, we were shortly to see.

* * *

Victory at Winchester

At this point a difference between the Winchester and the Aire Valley struggle requires to be made clear. The aim of the M3 Joint Action Group was not the abandonment of the inquiry, but its transformation to incorporate a re-examination of the question of need – a re-hearing of the Line Order Inquiry in other words – so that their plans for the improvement of the existing A33 could be properly examined as an adequate alternative which would minimise both cost and environmental damage.

I had no wish for the Department to have advance notice of the strength of the opposition. As it had been agreed that I would not act for the M3 Joint Action Group itself, but for individual objectors, I was able to lay something of a false trail. I was reported in the *New Civil Engineer* as having failed to get the support of the M3 group and I gave the impression of considerable disappointment. This, I believe, produced interesting results in the form of two road lobby bully boys (both 'respectable' and senior officials) who attended on the opening day specifically to shout me down. One, as reported in the *Observer*, white-faced, screamed abuse at me. In fact, he committed open assault, though not battery (he stayed his hand within millimetres of my face). Just before the inspector's entry one of them shouted at that packed Guildhall: 'Come on, you people of Winchester, what is this man John Tyme doing here?' to receive a howl of anger from hundreds of throats. Thereafter these two gentlemen (one of whom was dressed like a commando) were closely confined in a corner by people concerned to keep the peace. But this is to anticipate.

The Joint Action Group had engaged Associated Planning Consultants to present their case for the improvement of the (already dual-carriageway) A33. I met Leslie Ginsberg, head of the firm, in his office early in May and we had a useful talk.

'But what,' I remember him asking, 'if the Major General won't allow need to be re-considered?'

'He will,' I replied.

He grinned. Perhaps he believed me. I don't know.

I made numerous visits to the city at weekends, meeting people of all kinds and in all sorts of places. There was remarkable unity across the whole political spectrum, though the great majority were, as was to be expected, High Tory and High Church, and many had distinguished careers behind them, in industry, the services, the law or the church. Then there were the young, including the boys of the college, whose *spontaneous* support was a welcome leavening. A great many determined and admirable women played key roles in the immense organisational task. To persuade so many people of such conservative, traditional stamp that they should take a lesson from the Aire Valley was an interesting task. That success was so high is a testament first to their fierce determination to preserve that city and

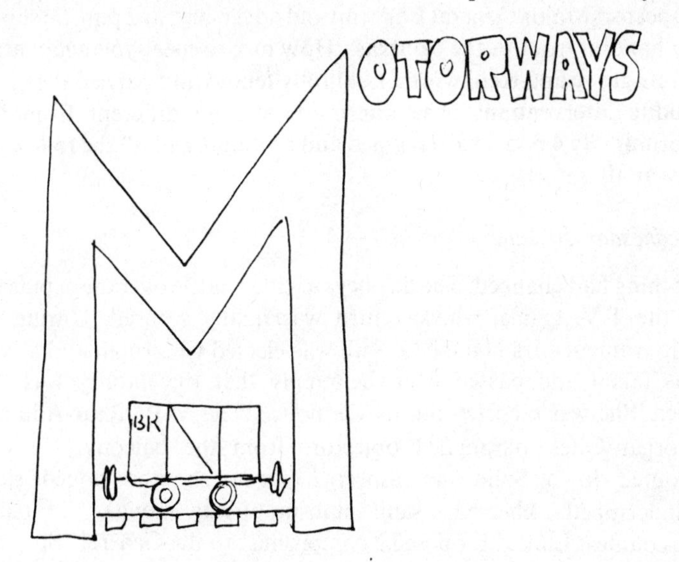

those watermeadows, and later to their conviction that only by resolute action could they defend what they came to see as democracy under threat. The unity and sense of common purpose was, as more than one person mentioned to me, reminiscent of wartime. An appropriate, carefully worded letter of authorisation was worked out, ultimately to be signed by no less than 518 objectors, including barristers, solicitors, councillors, retired service officers, etc. The small meetings throughout the town and the surrounding villages were organised by a massive telephone operation led by Pat Loveday, my host. They culminated in three crucial meetings in a room at the Indian Arms which later served as press headquarters. It was well named, I thought, for by now that remarkable summer was well under way and the temperature at those meetings put one in mind of the Black Hole of Calcutta.

* * *

Tuesday 29 June

Around 800 people present. The T.V. were there in strength. So were the police. The first person to be escorted out by the police was a city councillor. The whole morning was uproar, intervention, debate, adjournment, disintegration. The battle was to ensure that I was heard before the inquiry could be regarded as officially opened. The Inspector, Major General Edge (my old adversary; at 2 p.m. he shook my hand in front of the cameras: 'How nice to meet you again after Bromsgrove and Kenilworth'), skilfully fenced and parried the often erudite interventions. The afternoon was no different from the morning. By 4 p.m. chaos reigned and he adjourned till the following day at 10 a.m.

Wednesday 30 June

Nothing had changed. The day began with a battle over the expulsion of the T.V. teams, whose return was finally secured. During an adjournment Mrs Natalie O'Neill was elected Chairman and a vote was taken and passed overwhelmingly that the inquiry was not open. She was escorted out by the police. The M.P., Rear-Admiral Morgan-Giles, harangued objectors from the balcony. He was shouted down. Solid and respected citizens shouted, booed, slow handclapped, cheered, sang anti-motorway songs . . . Finally, Viscountess Enfield proposed a compromise to the General: 'Suspend a decision as to whether the inquiry is open or not and, "without prejudice" hear Mr Tyme.' The inspector agreed to hear me at 2 p.m.

At that hour I commenced the reading of the submission. This (see Appendix 3), vital to any understanding of the issues, effectively said the following.

First, that the Minister's statements in 1973 and 1975 that there was an 'urgent need for' the M3 Winchester section were both absurd and mendacious when set against the admissions of the Transport Consultation Document (H.M.S.O., 1976), for example:

'Yet by common consent, we still lack a coherent national transport policy.'

'By far the most important [objective] is the need to clarify the precise objectives of a national transport policy.'

The quotations made it clear that he could not either in 1973 or 1975 or any other time say there was an urgent need for anything, let alone a six-lane highway running around Winchester. Accordingly, my clients did not intend to see their properties destroyed and their environment wrecked for a scheme informed by such mendacity . . .

Second, that by the same token the Minister could not have complied with the 1959 Highways Act in the matter of local and national planning.

Third, that the statement issued to objectors claiming that the 'size and shape of the roads programme are . . . subject to the control of Parliament' was equally mendacious. I showed by references to Hansard and in particular to the debate of 1 May 1975 that far from controlling the roads programme, Parliament knew nothing whatever about it, being consistently denied information.

Fourth, that the Department had failed to comply with a requirement stated in the House of Lords by Baroness Stedman, whereby they were required to state the 'general background of national transport policy.'

Fifth, that the Department had failed to comply with the Council on Tribunals' interpretation of its obligations.

Sixth, that, in that the Minister had admitted the inadequacy of his road inquiry procedures following the Aire Valley battle, this inquiry should be adjourned until procedures had been rendered adequate.

Seventh, (and all-important), that at the 1971 Line Order Inquiry objectors had been denied the right to object to the motorway on the grounds of need; accordingly, the inquiry should be adjourned, and the matter of need should be the subject of a fresh inquiry.

At this point, I had to deal with a problem. The 1971 Highways Act incorporates an appeal procedure to the courts, but restricts the period to only six weeks. Winchester objectors were clearly outside that statutory period and thus could be said to have lost their case by default. The Ex Parte Ostler Appeal Court ruling (see chapter 4) emphasised this and made it necessary for me to appeal to the inspector on grounds of natural justice. There were five grounds as set out in full in Appendix 3 (pp. 149–53).

It is important to realise that this, the reading of the submission, was victory in the first skirmish only; the real battle lay ahead. Rebuttal by Counsel for the Department followed, followed in turn by my counter-rebuttal. Then points of order, speeches, accompanied by cheers and clapping followed, with arguments of considerable complexity from the floor. Amidst it all there was an interesting intervention. A man stood up in the balcony and in a commanding voice said:

'I hadn't read Mr Tyme's submission, and I didn't authorise him to speak on my behalf.'

He paused – what was coming?

'But there must be thousand upon thousand of people in this area who support every word he says.'

There were cheers and applause. And so the afternoon wore on with repeated demands for the abandonment of the inquiry. These were countered by Major General Edge repeating over and over again that, unless entirely convinced of the illegality of the proceedings, it was not in his power so to order. But the relentless pressure gave every indication of increasing and at the end of the day he agreed to give his ruling the following morning.

Thursday 1 July

The key words of the ruling were as follows:

I cannot disregard objections to the original Line Order. It would be unlawful if I did. I do not intend in any case to disregard objections to the Line Order rather than Orders for this Inquiry. I will hear all relevant evidence on the subject. Although it is premature to say this, I will listen to evidence on the need for the motorway. I do this for reasons Mr Tyme covered in his excellent submission; because the original inquiry was five years ago and much has changed since then – legislation, public attitudes, other things; it would be contrary to natural justice to exclude evidence that should have been heard in the 1971 Inquiry . . .

(As taken down by a trainee barrister)

This was a victory, but the battle was still very far from won. *'I will hear objections to the Line Order' can mean anything – or nothing. If the Inspector was to hear and consider objections properly, then there was need for a substantial adjournment.* And this, naturally, he was unwilling to grant. Counsel for the County Council saw no reason for one (which is incomprehensible until one remembers that Road

Construction Units, organs of Central Government, have totally destroyed any independence once enjoyed by county council transport departments). But Winchester City Council was a different matter. Their counsel rose and said:

> I was instructed on behalf of my clients to attend an inquiry into Compulsory Purchase Orders and Side-Road Orders. If this Inquiry is now to consider the need for the motorway, I shall require further instructions. For this my clients will require an adjournment.

This was something the inspector could not ignore, and thus at lunchtime the inquiry was adjourned till Tuesday 13 July. The battle was still very far from over, but this was a respite, and the scene was now to shift to a theatre on the outskirts of the town.

That lunchtime I bought a pint for David Keene, who, having acted for the Department at the Aire Valley, here was acting for the M3 Action Group and Winchester College. He should, now I come to think of it, have bought me one.

13 July

Heavy police attendance was in evidence. I had been authorised to read a second submission beginning with the words:

> Sir, it is now the view of many of my clients that the real issue before us is no longer that of a road, with the need for it or otherwise, or even with transport planning or the lack of it. They now see these proceedings as no less than an attack upon the rights and freedoms of citizens under the law, and they believe that the issue can best be expressed in the form of this question: Which is to prevail in this room today: The rule of law and Parliamentary democracy – or centralised bureaucratic power, responsible to no one?

I was refused, but persisted, and was escorted out by the police. But this was for the moment of no consequence as the issue was now out of my hands and in the hands of the lawyers, chiefly David Keene, and John Spokes Q.C. who acted as an individual, but in support of the objectors. The vital objective (without which nothing) was to ensure that consideration of need was a reality, and that required two things: production of the Department's full case justifying need, including their cost-benefit analysis (C.O.B.A.) *and* the M3 Joint Action Group's case, rendered in full C.O.B.A. form (the latter would take several months to prepare).

As I waited outside in the vestibule under the watchful eyes of the police I received messages to the effect that the inspector was obdurate; momentum was being lost. It seemed clear that a reassertion of 'popular power' was required. So, seizing a suitable moment when police attention wandered, I re-entered the theatre, raced down the aisle and faced the general eyeball to eyeball. The effect was dramatic. Two masses of people moved towards me at once: the men of Winchester and the police. The resulting mêlée, with scores of people, including the Headmaster of Winchester College, escorted out, fully restored the collapsing situation. It was a reminder to the General of the power of the forces still ranged against him.

According to Ann Morrow in the *Daily Telegraph* of 14 July:

> There were wild and chaotic scenes at the resumed inquiry into the proposed M3 at Winchester yesterday when 100 protesters were led away by strong-arm stewards and policemen . . .
>
> The protesters were not young with long hair and sawn-off jeans but teachers, doctors, clergymen and their wives. Among them was Mr. John Thorn, Headmaster of Winchester College . . .
>
> The hearing got off to a sober start with a warning by [Major General Edge] that he would not tolerate any interruptions. At the previous hearing there was uproar and it was then decided there should be an adjournment to yesterday.
>
> There were shouts, handclaps and stamping of feet and cries of 'Let Mr Tyme speak' as Mr John Tyme, senior lecturer in environmental studies at Sheffield Polytechnic, rushed into the theatre and tried to speak . . .

Eventually he was forcibly ejected by the stewards, who said they had been recruited from the Road Construction Unit of the Department of the Environment.

As Mr Tyme fumed outside the theatre, Mr David Croker read a statement on his behalf: 'It is my brief now to make it clear to you that by filling this hall with policemen or seeking to intimidate people under the Public Meetings Act or any other thing whatever, you do not weaken my clients' resolve.'

At this point, Mr Croker said with a little sarcasm to the inspector: 'You seem to have done very well this morning in this respect' . . .

In the afternoon the hearing became a shambles as the audience turned their backs on the inspector and sang from hymn sheets 'Rule Britannia' and 'Land of Hope and Glory' with slight adjustments like the second verse: 'Sewer of pollution, motorway to be, how shall we forestall three [M3]?'

More than a dozen burly policemen removed the protesters who had refused to sit down and be quiet . . .

There was a roar of approval as the Headmaster of Winchester College, Mr John Thorn, marched into the hearing but as he was escorted out he shouted: 'I'll be back.'

At the end of the day, the protesters seemed no nearer securing the adjournment they desperately wanted.

The inquiry was adjourned to the following day.

14 July

'What is going to happen if he still won't give way?' I was asked by a leading barrister.

'Have no fear,' I answered. For preparations had been made. If the final concessions were not to be made this day, then Winchester was to take the gloves off. So far people had agreed to leave when requested by the police; they would agree no longer, but would require to be escorted or carried out. It would be an Aire Valley par excellence. Other plans had been laid, and together would have led to very real trouble and difficulty. For as someone in authority remarked to me: 'We can't treat Winchester people like football hooligans.' At two minutes to ten a prayer was read by the wife of the Dean.

But before any drastic measures were needed, the inspector made the final concessions. In his ruling he acknowledged that need could only properly be ascertained 'by a comparison of equals' and that the cost-benefit analysis of the Joint Action Group would take a considerable time; insisting that the unwilling Department co-operate in its preparation, he adjourned the inquiry until 14 September. This was total victory. *The Line Order Inquiry was effectively re-convened. The battle was won and Winchester objectors could now exercise their right under the Act, denied to them in 1971. Thus did the balanced equation make that which was impossible at Hornchurch possible at Winchester. The Secretary of State HAD signed the Line Order. It WAS made. It WAS a fact of law. BUT IT WAS UNMADE AND THE ISSUE WAS RE-OPENED.*

* * *

In fact, the inquiry did not re-convene until 2 November, and in total it was in session for 111 days. During that time, at a cost to the Action Group of approximately £40,000, the following matters were established by their very experienced and highly qualified professional team of consultants:

1. The motorway could not be built unless it was proposed to abandon all pretence of following the Department's own technical memoranda;
2. The rate of return, on their own terms, was found to be less than the required 10 per cent;
3. The cost-benefit analysis was incorrectly applied, giving an entirely false and exaggerated result;
4. There was over-production of heavy goods vehicles;
5. The growth of population was over-estimated by 15 per cent;
6. The £20m cost was a major under-estimation; £30m at 1975 prices is a more realistic figure;
7. At no stage are the Department's figures subject to independent audit.

In sum: no examination of need had ever been undertaken. There is, in fact, no need for that section of the M3 motorway. The improvement of the existing four-lane highway would satisfy all requirements.

4

Chichester to Boston: the Concept of Natural Justice

The principle 'audi alteram partem' goes back many centuries in our law and appears in a multitude of judgements of judges of the highest authority. In modern times, opinions have sometimes been expressed to the effect that natural justice is so vague as to be practically meaningless. But I would regard these as tainted by the perennial fallacy that because something cannot be cut and dried or nicely weighed or measured, therefore it does not exist.

Lord Reid (*Ridge* v. *Baldwin*, 1964)

Chichester: the Conundrum

One important requirement of natural justice is that the objector should have the opportunity to know and meet the case against him.

Professor Wade, *Administrative Law*

The letter from the Dean on behalf of the Chichester Society arrived in the middle of the Winchester inquiry and when that was over I visited the town and met the Committee. It was an unusual case. The proposal was to render permanent a six-month trial pedestrianisation of the centre streets of the town. The objectors, consisting of thousands of concerned people, contended that, though superficially attractive, the proposal had hidden implications relating to the damage from juggernauts to certain attractive and historic buildings along the rear access roads in certain areas of the town. It was the hidden implications and what lay behind it all that made me keen to take the case.

When it became public knowledge in the town that I had been engaged (for my usual fee of 50p), certain people conducted a campaign of personal abuse, including the distribution of 5,000 leaflets headed 'Tyme Out!', claiming that I was a threat to law and order in the town. As a result of this the local Chamber of Commerce withdrew their support, but nevertheless, when I rose I was able to state that I was acting for something approaching 4000 people.

I opened my case by establishing the existence of the principle of natural justice with the extract from Lord Reid's 1964 judgement which heads this chapter and proceeded to read the following extracts from the authorities and commentaries:

1. The right to know the opposing case. One important requirement of natural justice is that the objector should have the opportunity to know and meet the case against him.
2. A hearing where the party does not know the case he has to meet is no hearing at all.
3. Ministerial instructions, therefore, ask local authorities to prepare written statements setting out the reasons for their proposals and to make these available to objectors in good time before the inquiry. (p. 233)

Wade, *Administrative Law*, 3rd edn (Oxford University Press, 1971)

4. Exceptionally, a duty to act judicially in accordance with the rule [audi alteram partem] will be held to arise merely by virtue of the impact of an act or decision on individual interests . . . (p. 162)
5. Protection from Jeopardy – The courts tend to lean in favour of according procedural protection to certain classes of individual interests when these are placed in jeopardy. These interests include . . . having one's property destroyed, taken away or substantially interfered with . . . (p. 162)
6. Yet it is sometimes patently unfair to deny a person in such a position the protection afforded by the audi alteram partem rule. And, as we have already seen, the exercise of wide discretionary powers in relation to private property rights may be held invalid for failure to observe natural justice, although the powers are typically administrative. (p. 169)
7. Prior Notice – Natural justice generally requires that persons liable to be generally affected by proposed administrative acts, decisions or proceedings be given adequate notice of what is proposed so that they may be in a position: (a) to make representations . . . (p. 180/1).
8. Duty of Adequate Disclosure – If relevant evidential material is not disclosed at all to a party who is potentially prejudiced by it, there is prima facie a breach of natural justice irrespective of whether the material in question arose before, during or after the hearing. (p. 191)

De Smith, *Judicial Review of Administrative Action*, 2nd edn (Stevens & Sons, 1968)

9. Time and again in the cases I have cited it has been stated that a decision given without regard to the principles of natural justice is void.

Lord Reid (*Ridge* v. *Baldwin*) p. 131

10. (Having quoted the United States Supreme Court) It can now, however, be confidently stated that even in our legal system the party concerned must be timely informed of the substance of the case he has to meet, which must be reasonably clearly formulated.

Sloan v. *General Medical Council* (1970), p. 120

11. . . . and it is also clear that if property rights are at stake, natural justice must be observed . . . (p. 127)
12. . . . it should be noted that the principles of natural justice are concerned with procedural matters and not matters of substance (p. 132)
13. It is now well settled that a statutory body which is entrusted by statute with a discretion must act fairly . . .

Breen v. *Amalgamated Engineering Union (ante) Lord Denning* (dissenting), p. 130
Garner, *Administrative Law*, 4th edn (Butterworth, 1974)

14. What then does fairness mean? . . . It is a tribute to the common law that the perimeter of the audi alteram partem rule has been enlarged, but, without some understanding of its contents, much of the ground gained in RIDGE may be threatened. It is towards this ambitious end that this paper has been directed. Put bluntly, fairness is nothing but natural justice: it is another name for the audi alteram rule. If there is any difference, which is strongly disputed, such difference must be trivial, for fairness lies within not outside the audi alteram rule. As to the elements of fairness, the following principles emerge from the cases:
(1) the person aggrieved must be informed of the gist (and not the meticulous details) of the case against him. (pp. 255–7)

C. P. Seepersand, 'Fairness and Audi Alteram Partem', *Public Law* (Autumn 1975)

I then proceeded to the matter in hand, stating the three grounds upon which my clients had been denied natural justice, these being: *first*, the fact that the decision regarding permanent pedestrianisation had been made by a body – the Highways Committee – upon which they had no proper representation and no hearing could not but be a denial of natural justice. *Second*, the fact that the proposal substantially prejudicing and injuring many of my clients had been made without any consideration or experimentation having been given to alternative proposals which would achieve the same ends without the said injury and prejudice could only be a denial of natural justice. And *third*, and most importantly, that natural justice was denied when a planning proposal was made without the real purpose of the proposal and the relevant evidential material being disclosed.

The argument regarding the third issue was then developed in detail. Why, the first question went, were the residents of Little London, the Pallents, etc. suffering from the ravages of juggernauts? To which the only answer was: because most of our freight travels by road and it is uneconomic to use small lorries. And the second question was: If the *status quo* was to be made permanent, was it not the case that either one of two things must happen: either the buildings would fall down from the reverberation and consequent structural damage *or* they would be demolished by planning order to provide better lorry access? And the answer to that question could only be: Yes. And the third and final question was: It surely could not be the intention of either of the planning authorites that this should happen, they having stated their objective in the Progress Report as being: 'The preservation of scale and character and, in particular, the safeguarding of buildings of special architectural or historical interest, together with groups of buildings . . .'

Thus, the conundrum, which, I suggested, could only be solved by there being some other matter so far unrevealed. And that so-far-unrevealed matter could only be, I submitted: very substantial roadworks, including the building of the M27 Chichester section and links to the town, together with a juggernaut park or out-of-town goods depot somewhere in the vicinity. In which case, the conundrum was answered. The residents of Little London, etc. would no longer have juggernauts passing their houses, but small lorries and delivery vans – and all would be well. Except, that was, for my clients who, I concluded, in the matter of that inquiry would have been denied the *audi alteram partem* rule in that the manner in which the proposal had been made had failed to disclose vital and all-important 'relevant evidential material'. Quoting finally once again from Lord Reid:

> Time and again . . . a decision without regard to the principles of natural justice is void.

I requested the adjournment of the proceedings so that the inquiry could be re-convened under circumstances where in all these instances natural justice would be observed.

* * *

It will come as no surprise that the inspector did not adjourn, and that his report and recommendations have left the town with a 'compromise'. Goods vehicles can deliver in the central streets overnight. It is hoped to relieve the intolerable conditions in Little London, etc. by traffic management schemes, a link road, a north–south by-pass and a juggernaut park. Meanwhile, the 'preferred route' for the M27 has been announced. Those roads, that juggernaut park and that motorway/trunk road constituted the real case that objectors should have been able to meet; the pedestrianisation proposal was merely a part of the case. Natural justice was clearly denied.

For me the Chichester inquiry created a personal crisis. Behind all that has been written so far lies an unanswered question: How did I manage to attend so many inquiries while employed as full-time senior lecturer at Sheffield Polytechnic? The answer is two-fold. First, I had to work harder. Final-year degree students can be trusted to work and during my absences either they saw films comprising part of the syllabus, or they were at work on their individual projects. My work in no way suffered for a further reason: students undoubtedly benefit from tuition from someone actively engaged in his own sphere and subject to the cut and thrust of the outside world. The second answer lies in the courageous and independent-minded support I received from my Head of Department and Principal. Their attitude was: As long as the work is done, we have no complaint. That the work was always done was invariably revealed at Examination Board meetings and Course Review Committees.

However, with Hornchurch following the Aire Valley and Winchester hard on its heels, pressure came to bear from powerful sources. Reference to my activities had been made more than once in both Houses of Parliament, and Written Answers in Hansard for 12 July 1976 includes the following:

> *Rear-Admiral Morgan-Giles* asked the Secretary of State for the Environment in how many road planning inquiries arranged by his Department objections have been raised by Mr John Tyme, a senior lecturer at Sheffield Polytechnic.
> *Mr Marks*: 'Full information is not readily available, but Mr Tyme is known to have appeared at 13 inquiries into objections to trunk road proposals since 1973.'

I was well aware of this pressure and foresaw increasing restrictions. Chichester, occurring at the beginning of the academic year, posed very real problems, and for the first time I found myself in the proverbial hot water. It proved very hot indeed and I was made very clearly to understand that things would have to change. Ironically, the people who had distributed the 'Tyme Out!' leaflets around Chichester subsequently put the rumour about that for my appearance I had received a fee of £1000 from the Communist Party. Far from receiving anything of the kind, I had been deducted two days' pay.

My dilemma was now extreme and I had no idea how I was to resolve it. Then, out of the blue, one evening after I had addressed the Halifax Civic Society I was approached by a trustee of the Rowntree Social Service Trust. He said that if I required any help I should get in touch with the Trust's Executive Director. The result was my present grant and my resignation from Sheffield Polytechnic. The grant is for two years only and represents a 30 per cent drop in my net salary. This has brought decided problems, particularly as a result of living and working in London. However, this *deus ex machina* has solved my dilemma for the moment and has enabled much to be done which would almost certainly have had to remain undone.

Belfast: Community Murder

Yet the defenced city shall be desolate, and the habitation foresaken, and left like a wilderness.

Isaiah 27:10

Late in April 1977 I was invited to Belfast by the Community Groups Action Committee on Transport. The Community Groups are from both Catholic and Protestant areas and constitute one of the small signs of hope in that city and province of desolation and despair. They formed their Action Committee in response to a typical Department of Transport/road lobby threat to their communities. The building of two urban motorways had already led to the destruction of approaching 4000 homes. Now they were threatened with a plan to link them up which, together with associated road schemes, would result in the demolition of some 2000 more.

It was my first visit to Belfast since the war. One afternoon and one night I spent crossing and recrossing the Peace Line meeting the members of the community groups at their various headquarters. It was, despite all the T.V. programmes I have seen, a shock of nightmare proportions. It was like a German or Russian city in the early months of 1945. Huge areas were rased; street after street empty, boarded up with corrugated iron; road blocks, barbed wire were everywhere; and over all there hung an air of poverty and depression such as the north of England knew only in the 1930s. This, of course, was and is meat and drink to the Department of Transport/road lobby. Here were no privileged middle-class communities who could defend themselves. Here was a demoralised, poverty-stricken, divided people. Six-lane highways, split-level junctions and slip roads could be imposed on these wretched communities with ease. There were fortunes to be made by demolition contractors, civil engineering consortia, purveyors of trucks, cars, road materials and petrol. And, by God, the Department of the Environment at Stormont, in closest co-operation with the Whitehall road lobby, was set to ensure that no legal impediments should stand in the way.

Thus was set up an elaborate farce known as the Transportation Study for Belfast, and this was the basis of the forthcoming inquiry. It conceived of the problem in terms of road-building; it set out a number of alternatives based upon road-building; it specifically excluded the one alternative that was not based upon road-building;

and it was carried out by a firm whose whole history was 'consultancy' on road-building.

If ever I had needed evidence of the mindless cruelty, the murderous effect upon helpless communities that is the lifetime achievement of those great highway mandarins who sit planning it all in Whitehall in conjunction with their friends in the road lobby, it was here.

* * *

The Community Groups, whose whole *raison d'être* is to heal the rift between Catholic and Protestant rightly saw this threat to both communities as a means of bringing them together, and establishing the beginnings of understanding and co-operation. As a result of our meetings, I was asked to return and act on their behalf at the forthcoming inquiry. It proved the most exhausting and harrowing of all, but it conferred upon me a privilege and honour beyond any that I had undertaken.

The relevant sections of the submission are as follows:

Sir, may I admit at my opening that I am not trained in law, and this I regret, as I do not share the distaste in which many hold the profession. When law, Sir, is allied with justice, it seems the most noble of entities; and it must now, *particularly in this province of the United Kingdom*, be clear to all of us that we are separated from chaos or worse by a very narrow barrier, a barrier no thicker than the paper upon which that justice incorporated with law is written. In this context, Sir, I might make the point that if people had ever taken the trouble to find out, they would know that every time I have been upon my feet at public inquiries or doing whatsoever I have done therein, it has been to seek that the law be upheld, together with those democratic rights and privileges which are our precious inheritance, and which are, in so many ways (largely from massive bureaucracies responsible to no-one, least of all Parliament), now under such threat.

Secondly, I should like to say that if what I have to say to you appears somewhat lengthy, I request you to accept my assurance that this submission contains not one wasted word. Furthermore, and most importantly, if what I have to say to you appears at first to apply to England and not to this Province, again I would beg your patience, for although this may appear to be the case, Sir, I shall not fail in the end to make what I have to say highly relevant to the matter before you, Sir, here in this Province and in Belfast.

I first require to raise before you the manner in which the decision has been made regarding this particular strategy option, and here at once, for my case to make any sense, I must depart for a while from this Province to look at the decision-making process in the matter of transport as it relates to England.

[Here follows much of the substance of the first Archway Road submission (see p. 65).]

And now, Sir, at last, may I bring us back, in this context, to this Province, to Belfast, and the subject of this Inquiry before you, Sir. With respect, it appears to my clients that the manner in which this matter has been dealt with exemplifies to perfection the corruption of function that I have, in its wider DoE London context, so far dealt with. They note, Sir, that the firm of Travers Morgan & Co. has been producing Belfast's transportation reports for over a decade, which in itself may be no bad thing, but they note from those reports one singular fact: in each case, the recommendation is for a solution requiring the building of massive road schemes. They deny, Sir, that Travers Morgan can be considered genuinely independent transportation consultants, not because of any improper action or intention on their part, but because their whole history and transport consultancy activity has been *within the ambit of the road construction industry*.

And therefore, Sir, when they see that the Department here at Stormont, having according to their Para 1.02 of their *Review of Transportation Strategy*, accepted that certain factors have changed since the Public Inquiry of 1972 'together with an appreciable change in public attitudes to urban motorways and their effect on the environment which has led the Government to decide to review the overall strategy on which the Transportation Plan has been based' – having accepted this, when the Department then returns the matter for reconsideration to the same road-building oriented firm which produced the original proposals; and when they note that this same firm now recommends the exclusion of the one strategy which by its acceptance for consideration would alone have indicated any genuine change in government's intentions, under these circumstances, Sir, my clients wish me to make it clear to you that they have *no confidence whatsoever in the decision-making process that has brought us to this Inquiry*. Now, Sir, I should not like you to under-estimate in any way the gravity of what I have just said. When people, anywhere in this United Kingdom, experience such a sense of betrayal, it is a matter of the utmost seriousness. It is an axiom, Sir, well known everywhere – and it must be particularly well known here – that *government rests upon the consent of the governed*.

I have referred to equity and natural justice. Allow me to explain in what manner the nature of this inquiry is a denial of both. Equity, Sir, 'equi', a prefix implying equality. 'Equity: right as founded on the laws of nature: moral justice, of which laws are the imperfect expression; the spirit of justice which enables us to interpret laws rightly; fairness . . .' (*Chambers 20th Century Dictionary*). Now, Sir, you hold at this moment an honourable trust. If you were an inspector at a similar inquiry in England, you would be in receipt of a document entitled *Notes for the Guidance of Panel Inspectors*. And there in, Sir, Para 1.13 would accord you this resounding title: the Custodian of Natural Justice.

Now Sir, it is my submission to you, that the manner in which the decision to hold this Inquiry upon its present restrictive terms is a total denial of anything relating to 'equality' or 'fairness'. For these concepts to have been incorporated, I submit to you that *both* of the fundamental alternatives (a pro-road and a pro-public transport alternative) should have been incorporated within the scope of this inquiry. Now, Sir, I am aware (but I hope that I shall be corrected if I am wrong) that the Tribunals and Inquiries Act of 1971 does not apply in this Province, which means that the body reaffirmed by that Act, the Council on Tribunals (to which all persons in England who feel that in any way an inquiry has been improperly held, or has in any way denied them their rights, may have recourse and seek redress) has no standing here. (I hope, in passing Sir, that I am wrong in this matter, as it seems to me such a fundamental denial of a form of justice to people of this province which is available to people in England, perhaps there is a similar body or remedy available here of which I am unaware: I have no doubt that either you, Sir, or Learned Counsel for the Department will be able to put the matter right here and now.) But in any case, to the point I am making. In the Council's interpretation of the all-important Franks Committee Report on Tribunal and Inquiry Practice of 1957, they lay great stress upon three concepts: openness, fairness and impartiality. It is my submission to you, Sir, on behalf of my clients that the manner in which the terms of reference of this Inquiry have been decided upon is a denial to them of all of these concepts. There has been inadequate *openness* (about which I shall be speaking in a moment): that being so, there is a denial of *fairness*; and in that the proposed strategies deny the real option, there is an obvious denial of *impartiality*.

I have referred to inadequate openness, and I wish now to move onto the matter of Natural Justice, of which you are, Sir, the Custodian and to that great pillar thereof enshrined in the Latin 'Audi alteram partem', 'Let the other side be heard'. I have referred to Chambers for a definition of 'equity'. I think it would be more appropriate to turn to the world of jurisprudence regarding 'Natural Justice'. [Here follow abstracts as quoted at Chichester (see pp. 44–5).]

And now, Sir, may I state as briefly as possible, the manner in which my clients – or some of them – claim that in the matter of 'audi alteram partem' they are denied natural justice. Sir, from an experience, which can now be said to be reasonably wide, of road proposals and public inquiries into them there is one thing that to me stands out very clearly as common to all of them, and it is this: at no time in my experience has the Department published the full implications in terms of subsequent road works of the proposals which they submit to public inquiry. Over and over again, it is discovered, subsequent to a given inquiry, that there are additional road works which are found to affect many people who, *for the vital period between the original scheme which does not directly affect them and the subsequent schemes that do*, find that objections to the latter have been totally prejudiced by the Secretary of State's decision on the former. The most common manifestation of this particular administrative practice is the dividing up of a line of motorway or trunk road into small portions, the completion of each of which subsequently prejudices more and more people entirely unaffected until draft orders are published for stretches that do directly affect them.

Now, Sir, what I am saying to you, in brief, is that the alternative strategies, which are the subject of this inquiry, themselves conceal subsequent road works in the form of link roads, junctions, road widening and improvement schemes which deny to those affected any knowledge of them but which, by the decision on one of the given strategies which are the subject of this inquiry, will then affect them in such a way as to deny them any means whatever of defending their property rights which, being at stake, entitle them to natural justice under the audi alteram partem rule as made clear by Garner. I could very easily find myself in this position in the district where I live, and so, Sir, with respect, could you. A draft line order could be proposed, leaving us entirely unaffected, but the departmentally-known subsequent schemes and improvements would not be disclosed to us, in which case we should come within de Smith's well-known interpretation regarding the Duty of Adequate Disclosure, the material arising 'before the hearing' would not be disclosed to us; we should be 'potentially prejudiced by it', and it would thus constitute, *prima facie*, a breach of natural justice.

Well, Sir, as *should* we, in that hypothetical case, so *do* my clients in this most decidedly non-hypothetical, and very real case which is the subject of this Inquiry. And so, Sir, what I am saying to you is that, as the Custodian of Natural Justice in these matters, and bearing in mind everything that I have said to you relating to the erosion of public confidence in everything relating to road planning, you should adjourn this Inquiry until everything relating to these plans has been disclosed. This is not asking an impossibility, but merely something well within the scope of road transport planning . . .

For, Sir, if ever there were anywhere on the face of the earth where natural justice needs not only to be done, but be seen to be done, it must be here in Belfast in this Year of Our Lord 1977. That is the weight of responsibility, Sir, that I place upon your shoulders when I beg of you, with every fibre of my being, not merely to adjourn this Inquiry pending an investigation of all the matters I have raised, but to require also the Department of the Environment here in Belfast to cease any and all road works and all vesting of properties, similarly pending that investigation.

There was no disruption, as I had been instructed by the Action Committee in no way so to involve myself. As the Belfast *Newsletter* reported I had 'kept the proceedings cool during the sometimes heated hearing in an Inquiry packed to the doors and charged with emotional atmosphere'. The disruption came after I had left. The inquiry was not adjourned, but a 'concession' was made. After its completion, three months would be given to the Community Groups (who have no money, of course, to purchase the necessary expertise) to produce their counter-proposals. Any more cynical tactic could hardly be imagined. On 8 June the inspector established a lock-out, so that the public inquiry could proceed in private. A word on the inspector, Mr Michael Lavery Q.C. I could not but respect and like him. Perhaps his decision to carry on was in the face of the obvious threat that should he resign his repellent commission, he would be immediately replaced on Whitehall's instructions by any one of those many to hand who cannot even know the meaning of honour and integrity, let alone observe them.

If ever there were anywhere on the face of the earth where Natural Justice needs not only to be done, but be seen to be done, it is Belfast in this year of Our Lord 1977.

Absurd of course to imagine that the road lobby mandarins of Marsham Street and their provincial lackeys could be deflected by such minor considerations.

The Ipswich Scandal

Statutory rules do not in general exclude the rules of natural justice, but are supplemental to them.

Sir Douglas Frank (*Nicholson* v. *Secretary of State for Energy and Another*, August 1977)

The road lobby regard the Ipswich by-pass as of major importance. They are right to do so; it plays a key role in their plan to drive an unspecified number of motorways and trunk roads across East Anglia and thus deprive British Rail of freight carriage between the eastern ports and the Midlands and the North. The inquiry into it, therefore, was a matter of great concern to them and the appointment as inspector of Mr Clinch of Epping fame must have been very reassuring, he having recently completed his report giving whole-hearted approval to driving a six-lane highway through what remains of Epping Forest.

The same road lobby official who had displayed such frenzied violence at Winchester was despatched to help things along; a major and sustained campaign was carried out in the local press; local firms were induced to believe, incredibly enough, that their future lay in road transport; on the eve of the inquiry drum majorettes marched through the town in support of the by-pass; and I am reliably informed that local firms gave their junior executives time off to attend the opening day. In the face of all this it behoved me to take precautions, and with objectors to the inquiry procedure in such a minority, it was agreed that I should on their behalf take it upon myself to make the necessary gesture, leaving them to remain in the inquiry free of impediment.

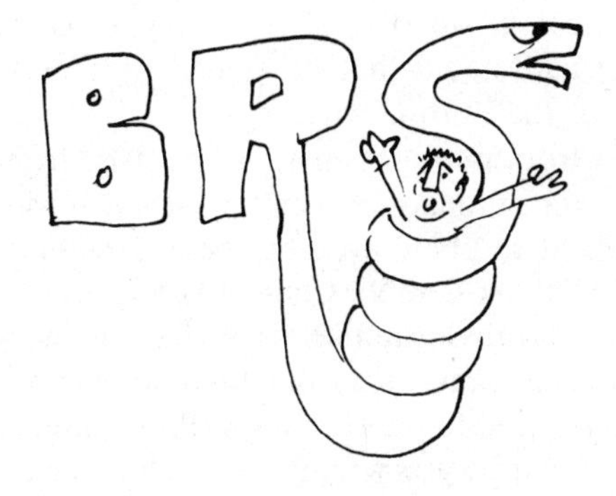

After the inspector had made his opening statement, I rose to say that I had a procedural matter to raise on behalf of over eighty objectors, both statutory and non-statutory, including the East Anglian branch of the Railway Invigoration Society. He ordered me to be seated, saying that I would be heard on that matter when I presented the objection. When I asked whether he was seriously suggesting that matters of procedure regarding the inquiry should be heard only towards the close of it, and reminded him that the *Rules for Guidance of Panel Inspectors* (a Department of the Environment document) appointed him 'custodian of natural justice', he interrupted me, saying that he took no note of this document; he was an 'independent inspector' and would hold the inquiry according to the rules laid down by the Lord Chancellor. Again he ordered me to be seated. I said that I was at his discretion in this matter and sat down. Had I not taken precautions, this would, of course, have been something of a defeat, for a major gesture was clearly necessary. Thus the chain by which I was able to padlock myself to a heavy table enabled me first to intervene, then to refuse to be silent and finally to be escorted out by the police, accompanied by the table.

The inquiry then proceeded along its arbitrary course, with objectors refused the right to cross-examine witnesses and required to present themselves at impossibly short notice, and with non-statutory objectors regarded as a mere nuisance, to be ignored apparently at will.

Some weeks later, I was asked again by the Railway Invigoration Society and others to act for them in a matter that had arisen regarding the conduct of the inquiry. Perhaps because the chamber was on that day empty of by-pass supporters, I was permitted to read a submission which recalled the opening day when he, the inspector, had appeared to take no note of the *Rules for Guidance of Panel Inspectors* and had stated that he abided by the Highways (Procedure) Rules of the Lord Chancellor.

I stated that from information I had received from my clients it appeared that in one fundamental matter he had, during the course of the inquiry, failed to abide even by these. I reminded him of the occasion when Suffolk County Council (who were statutory objectors) had wished to cross-examine Mr Wales, a witness for one of my clients, but had been prevented from doing so by him, the inspector, and that when referred to the Highways (Procedure) Rules, Para. 13 which indicated that he was wrong in his ruling, he had replied that this was not so; objectors could only cross-examine witnesses from the opposing side. I informed him that another of my clients had then written to the Department for guidance on the matter and had received a reply supporting the objectors' case.

I then referred him to the *Rules for Guidance* Para. 1.54, which reads:

> *Questions to Friendly Witnesses*
> . . . where Rules of Procedure apply, they provide that the promoting authority and the statutory objectors shall be entitled to cross-examine persons giving evidence and make no distinction between friendly and hostile witnesses . . .

saying that if there were any doubts after the letter from the Department, this must now dispel them.

Pointing out that the matter could not be said to end there, I said that Mr Jeremy Coles, one of my clients, had been denied the opportunity of cross-examining witnesses and in his proof of evidence he had made clear his frustrated wish to do so; and that that being so, the question must arise: how many other objectors had been similarly prejudiced – and that the answer to that question must be that we could not know.

I submitted that this must call into question the inspector's conduct of the whole inquiry. It had been characterised, I said, by a veneer of concern for the interest of objectors, which had obscured his arbitrary and capricious determination to deny them their plain rights under the Lord Chancellor's Rules. This must be seen, I submitted, as sufficient to require the quashing of the proceedings and the reconstitution of the inquiry under an inspector who would not only abide by the said Rules, but have the goodness, not to say the common sense, at least to *read* the *Rules for Guidance*, which were set out to ensure, *inter alia*, that the inspector was, as stated in Para. 1.13, the 'custodian of natural justice' whose duty it was 'to ensure that all parties have fair opportunity to present their case'. Finally I stated that as he had neither been such a custodian, nor ensured that fair opportunity, my clients intended to seek to have the proceedings quashed *ab initio*.

Counsel for the Department then rose and in rebuttal stated that Suffolk County Council had subsequently been given the opportunity of cross-examining the witness. I made no counter-rebuttal, but had I done so it would have been to say that that was nothing toward, and did nothing to alter the impression that objectors had taken away regarding cross-examination, which had in every case substantially prejudiced them.

The following month a judgement in the High Court (*Nicholson* v. *Secretary of State for Energy and Another*) pronounced very clearly on this matter:

> . . . it did not follow that, because a certain objector did not have an express right to cross-examine, he might not nevertheless have such a right as a rule of natural justice, depending, for instance on whether his objection was bona fide. Although the inspector had a discretion, that discretion could not be exercised to exclude the rights of non-statutory objectors if that exclusion would be contrary to the rules of natural justice. It was beyond question that statutory rules did not in general exclude the rules of natural justice, but were supplemental to them. (*Times Law Report*, 5 August 1977)

and on the request of Counsel for the Department the Ipswich inquiry was adjourned *sine die* so that witnesses could be contacted and enabled to cross-examine appropriate witnesses. To what extent that would have been possible within the time available is wholly conjectural. It was then reconvened and concluded in a matter of days. Thus ended the Ipswich scandal and Mr Clinch at the moment of writing is composing his report.

Boston and Magna Carta: the Ostler Case

Nullus liber homo . . . disseisiatur . . . nisi per legale judicium parium suorum vel per legem terre.

Magna Carta, Cap 39

The Ex Parte Ostler Appeal Court judgement was referred to in the Winchester submission. It is interesting on its own merits. A curious tale unfolded over the winter of 1976/7. It had begun in 1973, when a public inquiry had been held into a minor road scheme, part of the general wreckage known euphemistically as the Boston Inner Relief Road. Mr Sydney Ostler of a centuries-old family firm of corn merchants did not object to the proposals at that inquiry, nor did he attend. His property was not affected – or so he thought. But unknown to him some objectors to the proposals, including a major property holder in the neighbourhood, were assured by letter from the Department of the Environment offices in Nottingham that Craythorne Lane would be widened to provide necessary access. The family firm of Ostler owned a small store in Craythorne Lane. As a consequence of this letter the objectors agreed to withdraw their objections, and the Department then made an agreement under Section 57 of the 1971 Highways Act with the Borough Council. This transferred authority for the road works from the local authority to the Secretary of State and *denied any member of the public a statutory right of objection.* The public inquiry into the Side Roads Orders and Compulsory Purchase Orders took place in September 1973 with no reference to the letter referring to the Craythorne Lane proposal, and the only conclusion that can be drawn from this is that *the Minister was to be induced to make an order which his officers had no intention of carrying out as stated.*

Some time the following December someone knocked at Mr Ostler's door and asked what sort of a price they could agree on for the firm's offices in Market Place as they might be required for a road scheme. No mention at this time was made of the Craythorne Lane property as the official concerned was unaware of any such building owned by the firm. At first mystified, Mr Ostler at once consulted his solicitor and a local councillor. The implications for his Craythorne Lane property at once became apparent and from their investigations they very soon concluded that some sort of secret deal had been made. *This, however, was denied by all concerned.* A supplementary

Compulsory Purchase Order Inquiry was then convened to acquire the Ostler property in Craythorne Lane and the inspector ruled that Mr Ostler could not raise the matter of the alleged secret agreement as that was not relevant to the planning merits of the case he was considering.

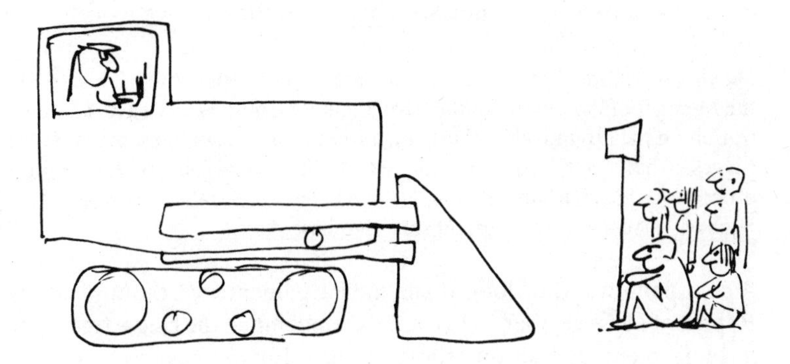

Meanwhile, the Side Roads Order, the subject of the September 1973 Inquiry, was made by the Secretary of State, and the inspector at the second Compulsory Purchase Order Inquiry found in favour of the order. In simple terms, nothing now stood between the Ostler property in Craythorne Lane and the bulldozer.

It was at this point that the intervention of the local M.P., Mr Richard Body, and a chance letter revealed incontestible evidence of the secret agreement. Mr Ostler at once took out a writ of certiorari to quash the orders made as a result of the 1973 Inquiry, and in the High Court a preliminary judgement was given in his favour. This did no less, but no more, than allow investigations to begin into the alleged fraud. But the Department then appealed on the grounds that Mr Ostler had failed to apply to the Court within six weeks of the making of the orders as required by the 1971 Highways Act. The Court upheld the Appeal and refused him application to the House of Lords. Two points require to be made here. Six weeks is an impossibly short time for complex matters to be worked out in detail and for the subsequent setting in motion of the necessary legal machinery. Secondly, and more relevantly, *how could Mr Ostler possibly have appealed within the statutory six weeks when evidence of the suspected fraud did not come to hand until months after the final date?* While upholding the 'out of time' Appeal, the Court expressed sympathy for Mr Ostler, and a suggestion was made that a remedy could lie in an action for civil fraud against the Department. But Mr Ostler was not interested. His concern was his building and his rights to protect it.

The Court's refusal of an application to the Lords opened up a new course of action. From notes prepared for the case we read:

I advise, therefore, that it is still competent for the complaint to be referred by the Member of Parliament to the Parliamentary Commissioner for Administration on the grounds of maladministration. I say this because the evidence suggests that Mr Ostler was deprived of the effective use of the Inquiry procedure which Parliament intended him to have under the provisions of the Highways Acts and the Compulsory Purchase Acts.

A complaint was duly lodged and the Parliamentary Commissioner's Report of 30 December 1976 roundly condemned the Department for misjudgement; for serious shortcomings; for *the incorrect use of a Section 57 Agreement which denied Mr Ostler the statutory right of objection which would have been his had an amendment to the published draft Side Roads Order been made*; and for failure to inform Mr Ostler of the Craythorne Lane scheme (the secret agreement) as early as was his entitlement. He made it clear that it was his opinion that *procedures laid down to ensure fairness and openness had not been carried out* and that doubts must remain as to whether all interests had received fair and equal consideration. Despite all this, however, he concluded that to no material extent had Mr Ostler been denied the 'effective use of the statutory inquiry procedure', and considered the Department's expression of deep regret and willingness to compensate his legal costs an appropriate outcome.

But Mr Ostler had a good, old-fashioned idea that law and justice are or should be one and the same thing. While by no means averse to receiving his costs, and regarding an apology as neither here nor there, he, a plain man, saw the matter in plain terms. The first inquiry was into proposals which prejudiced his property and he had been denied knowledge of it; something was fundamentally wrong and required to be righted. It was difficult to disagree; nothing, it seemed to me when I heard of the matter, better exemplified Professor Wade's interpretation of the audi alteram partem rule: 'A hearing where the party does not know the case he has to meet is no hearing at all.' The fact that Mr Ostler was not a party to the dispute because of a secret agreement compounded rather than negated the denial of justice.

By now local councillors had taken an interest, and an interview was sought with the Minister. But this was consistently refused over a period of months. When it seemed that the arrival of the bulldozers was imminent, Mr Ostler's solicitor contacted me, and the facts relating to what happened that January 1977 are as follows.

Considerable media publicity was given to my intervention, this because I had recently made it known on a national television news interview that site confrontations would follow when objectors had been denied their rights in law. Yorkshire T.V. even sent a helicopter to pick me up at Lincoln racecourse.

In the ensuing interview I made it clear that here was a man who, despite all the legal processes that had taken place, had been denied natural justice and, to the plain man, clearly prevented from exercising his true rights in defence of his property. I referred to that traditional bastion of liberty, Magna Carta, and to Cap 39, from which we read:

> Nullus liber homo . . . disseisiatur . . . nisi per legale judicium parium suorum *vel per legem terre.*
>
> No free man shall be deprived of his property except by the lawful judgement of his peers *or by the law of the land.*

Site confrontation, I said, would be supported from all over the country. Accordingly a large number of telephone calls were made. I remember it was frosty weather and the Archway Road people, whom I asked first, were willing but apprehensive. However, they had no need to worry. *Within hours of the Yorkshire T.V. interview Richard Body, M.P., rang Mr Ostler with the news that the Minister had now consented to see a deputation of councillors.* On 24 February 1977 the proposal to widen Craythorne Lane was withdrawn. Thus the 'disseizing' of Mr Ostler's property has now been quietly dropped, other methods of making the road scheme have now been found and the law of the land has been observed in true substance in the sense that it has not been set aside in the matter of Mr Ostler's property.

But hereby, of course, there hangs a tale. When, two years prior to these events, the Conservation Society contemplated going to the

Courts on the issue of the Department's failure to comply with the Act in the matter of the statutory public notice, it was advised that it could take no proceedings with any hope of success; English law relating to administrative matters, it was advised, was in a mess and needed thorough reform and rationalisation. The Ostler case would seem most powerfully to substantiate this. The present position would seem to be that there are no means under administrative law of rectifying injustices and improprieties which take place as a consequence of compliance with statutory requirements. So long as the law cannot remedy the law's abuse (particularly when that abuse is a weapon of departmental or bureaucratic power), and when it could be shown that the threat of site confrontation and national outcry alone could bring about the desired end in the case of the Ostler property, this nation can find little for its comfort.

5

The Battle of Archway Road

It is not that people are utterly selfish. During the war they were ready to make great sacrifices, because it was clear why they were being asked to do so. So, too, people now could be reconciled to sacrifices on their part if they were shown to be necessary. The trouble with the present planning procedures is that they do not, and as they are at present constituted cannot, achieve any measure of reconciliation because the onus of proof is the wrong way round. The damage done by the department's proposals is obvious, real and keenly felt; it is not obvious that there will be corresponding benefits – the department does not prove that there is any need for its proposals, only asserts that there is – nor is it apparent that the benefits, if they exist, will outweigh the losses.

J. R. Lucas, *Democracy and Participation* (Harmondsworth: Pelican Books, 1976) pp. 275–6

The Archway battle, like that of the Aire Valley, was concerned from the beginning with getting the inquiry – and thus the road widening scheme and its side roads – abandoned. It was, and at the time of writing still is, designated as a Side Roads Order Inquiry, the side roads being the result of a Line Order (road widening and realignment) Inquiry that had taken place in 1973.

Several things distinguish it from other campaigns. First, the almost indescribable hell that motorway-generated traffic – chiefly juggernauts – has created for some local residents has ensured a vociferous, if not numerous, lobby in favour of the road. I understand their anguish, but in that they are the helpless pawns in the brutalised campaign to shift freight from rail to road, and in that the road widening would simply push the problem onto others who would then become the next set of pawns, I have always sought to convince them that their problem could be and should be solved by other means. Secondly it concerns an urban motorway (or 'road built to motorway standards') and in that it would merely have the effect of creating a high-speed link between one traffic jam and another, it is in conflict with every utterance on urban roads made by the House of Commons Select Committee. Third, with its lengthy and numerous adjournments, it has become the longest road inquiry ever; at the moment of writing it is a year and three months since it was originally convened. Fourth, it has been characterised by singular bitterness, occasioned by the 'private chamber' (bunker) sessions.

It had one thing in common with Winchester, however. People had been utterly disillusioned by the 1973 Line Order Inquiry and the report of its inspector, Mr Clinch. At one time, people of the area chiefly affected by the construction of the side-roads had decided from their experience under Mr Clinch that there was no point even in attendance. But the Aire Valley and Winchester had changed all that. Demoralisation gave way to inspiration and determination in a matter of weeks. That it did so was due to the work of a group of remarkable men and women, chiefly the latter. The small meetings, known now as 'street meetings' were organised with great efficiency and I was able thereby to meet hundreds of people from a large number of neighbourhood organisations in front parlours and dining rooms to explain in detail what required to be done and how.

My role in the long battle has been sporadic, and from the beginning I was not alone. Professor Terence Morris, who had played a significant role at Winchester, joined me in the procedural attack. His submission (Appendix 4) covered all aspects but one, including a demand for a reconsideration of the need for the original road widening. I decided to concentrate upon one issue – the overwhelming and corrupting influence that the road lobby has within the Department of Transport. From the beginning, therefore, it was a joint effort and involved others apart from myself, as a piece in the *Observer* for 19 September 1976 makes clear:

> While two rows of Department of Transport officials sat staring gloomily at their draft orders and their traffic flow tables, the air was full of St Thomas Aquinas and Montesquieu . . .
>
> The most striking characteristic of the inquiry was the objectors' bitter disbelief in the open-mindedness and even-handedness of the road planners.
>
> Mr Tyme's value to them is not primarily his statement of their argument – many of them were as articulate as he was – nor even his tactical advice, but the assurance he gives to naturally conformist people by his palpable presence that they are not isolated . . .

My 'statement of their argument', for which I was authorised by over 600 signatures, was submitted on 15 September as follows:

Sir, in showing you that to open this Inquiry would be wholly improper and contrary to the national interest, I require to show you that the authority you have been granted to preside over it is itself improper in that it derives from a wholly corrupted process of government, and it is for this reason that with the greatest respect, I request you to listen to what I have to say, and to make a judgement upon it quite independently of these papers before you whereby your authority purportedly derives.

Sir, on a B.B.C. T.V. programme in March of this year, the then-chairman of British Rail, Sir Richard Marsh, was asked this question:

'What do you expect, Sir Richard, to come out of the Government's current transport review for your railway?'

It must have seemed a perfectly normal and innocuous question. But it received the astonishing reply:

'Well, I am like the public. I read my newspapers. It would be nice to know.'

Well, Sir, of the 560 objectors for whom I acted before Major General Edge at Winchester, five were barristers-at-law and a very large number were solicitors, and amongst the matters that I raised on their behalf was what I termed the corruption of the decision-making process within the Department whereby, due to close co-operation between civil servants of the Department and employees of the Road Lobby, what the nation receives (as exemplified by the proposal before us at this resumed inquiry) is not a transport plan in the national interest, but a roads and motorways plan in the road lobby interest, or to quote from the *Architects' Journal* editorial for the 25th August of this year: 'Apparently, when the DoE pronounces on transport, it speaks with the mouth of the road engineer.'

Sir, my clients here look at the self-evident ruin that is the transport system of this country today: they see a truncated, semi-bankrupt railway network, in freight terms massively under-used: they see a grossly neglected waterways system for which minimal funds are consistently denied by the Department; they see a local authority transport network which has left all those without access to cars (approximately 50 % of households) less mobile than they were in the 1920s; they see the Department now demanding the withdrawal of subsidies for the wretchedly minimal services that remain; they have heard repeatedly Departmental spokesmen say that there is no money for any of these things – and yet, for a road network that cannot in any sense be completed until the late 1990s they see hundreds of millions of pounds being made freely available, and here in this area on the Archway Road they see a concentrated brutalised hell of juggernaut traffic, and they see that to 'answer' this brutality, which is an offence against every civilised value, it is now proposed, at immense cost, to extend the roadways, and in the process knock down 150 dwellings involving some 200 families.

And they ask themselves, Sir, how has all this happened? And they require me to say to you, Sir, that the answer to this question, as I shall be putting to you in detail in a moment is why I shall be asking you to adjourn these proceedings forthwith. And they ask me to put before you, Sir, this significant fact: that while Sir Richard Marsh was having to read his newspapers to find out what was going on in the transport planning of the Department (and the British Waterways Chairman was similarly excluded from any involvement whatever in transport planning), the British Road Federation, in the person of its Director, Mr R. H. Phillipson, together with eleven members of that Federation, all of whose firms derived and derive financial benefit from the building of roads, were, while all these road networks were being planned, while British Rail and Waterways freight carriage was being run down, and that horrendous endless stream of juggernaut traffic was being built up on the Archway Road, those twelve people were sitting cosily inside the Department, together with a helpful coterie of 25 Departmental Highways Civil Servants, on the British National Committee of the Permanent International Association of Road Congresses, the stated objectives of which are to: 'Foster progress in the construction, improvement, maintenance, use and economic development of roads, and encourage the use of road systems throughout the world . . .'

And to quote from the authoritative work *Wheels within Wheels: A Study of the Road Lobby* (Hamer F.O.E. Ltd, 1974): 'Due to this organisation alone, the Road Lobby has regular contact with the Department. None of the opposing lobbies on transport enjoy a similar relationship with Government on an international level, neither is there a comparable body for any other mode of transport.'

Elsewhere in this work (p. 29) we read the following significant statement: 'The Lobby needs the help of the Department when administrative difficulties arise which affect its members' interests.'

Sir, from the evidence, my clients are sure that this is true; and almost certainly one of the 'administrative difficulties' that arose for the members of the British Road Federation was the possibility that British Rail's electrification plans for north and east London would substantially forestall their clear objective so to cram the Archway Road with juggernaut traffic that they would be enabled to demand the extension of the road widening further up the Archway Road (and thence, as the inexorable logic of the process requires, on further to include Falloden Way).

Well, Sir, in my submission to you, it has certainly had the help of the Department to get over this 'administrative difficulty'. And I quote here, Sir, from a letter to me from Mr Richard Hope, Editor of *Railway Gazette International*, dated 21 March 1975:

The railway from Barking down to Tilbury is already electrified on the

25 kV overhead system identical to that used north of London, but a gap of sixteen miles exists between Barking and Willesden Junction via Tottenham. This prevents the use of electric locomotives to haul freight to and from the Tilbury area, so that locomotives and crews must be changed at Willesden Junction. Based on the costs for recent electrification schemes, and including an element for resignalling to increase line capacity. I would estimate the cost of closing this gap at £3m.

The relationship between British Rail and the DoE over electrification is that Government has consistently refused to agree to any kind of programme such as exists for motorway construction. This results in much higher costs . . .

No electrification scheme of any consequence has been authorised since August 1971 . . .

The point I am making to you, Sir, is that the very real 'administrative difficulty' that would have been experienced by those road haulage contractors had British Rail been able to tender for electrified freight transit direct from Tilbury to the Midlands and on to Scotland (had it not been so conveniently overcome) would have removed at a stroke half of those juggernauts at this moment thundering up and down the Archway Road; and such British Rail tendering would, of course, similarly remove those prospects of lucrative contracts for road construction which other members of the British Road Federation are, no doubt, confidently looking to, were you to open and preside over this Inquiry. And so, Sir, in a word, the Lobby's 'regular contact with the Department' has got them over that 'administrative difficulty', and, no less, is the reason why we are here today.

If further evidence were needed of the plans to overcome this particular 'administrative difficulty' for the Lobby, it is to be found at Tilbury. On the

4th of this month I was shown a great acreage beside the Royal Albert Dock (which was crammed with ships at the time I was there), for which planning permission has already apparently been granted for the construction of a juggernaut lorry park. The implications for the people of the Archway Road, Falloden Way and all over London of that proposed construction would not, of course, be the subject of this inquiry were it to open. But the implications regarding the Lobby and its close relationship with those Civil Servants in the Highways Divisions of the Department are my present concern before you now, Sir. For its construction would, of course, remove any *future* 'administrative difficulties' that the Road Lobby might face in the form of rail or waterway competition by maintaining an even greater level of juggernaut traffic in and around London.

And so, Sir, to come to the point, what my clients are saying to you is that during the period in which all those road planning decisions were made (without any Parliamentary approval, of course, as I showed at Winchester) specifically designed to lead to overwhelming traffic, and in particular juggernaut traffic, in this area (and which as a consequence have led to this particular widening proposal) – during this period, the closest co-operation has existed between accountants, economists, statisticians and road planners, both from the Department and the British Road Federation, so much so that, to quote Mr Hamer again: 'Certainly, the British Road Federation sometimes gives the appearance of having acquired the status of a Government body in recent years.' Or to put it in plainer form, my clients submit to you, Sir, that like two mice at a piece of cheese, Departmental Highway Civil Servants and Road Lobby representatives have systematically over decades dismantled the transport system of the United Kingdom in that Lobby's interest; and that being so, this proposal before us to inquire into compulsory purchase orders relating to the widening of the Archway Road is nothing but the product of a Departmental process designed to subject the national interest to the commercial; and this being so, I submit, Sir, that these proceedings be terminated at once to await the large-scale and independent investigation regarding the actions of those individuals in the public service who have done their part to bring it about.

But, Sir, the matter must go beyond this, must it not. For do we not have a Parliament? Have we not our elected representatives in the Commons who pass statutes, authorise statutory instruments, delegate powers to Ministers and so forth? Have we not recourse to them to prevent all this, to bring about the necessary reform and restitution?

Sir, it is now my submission to you that the Road Lobby have ensured that M.P.s were supplied continuously with papers containing technical, financial, economic and statistical information regarding which they had no means of distinguishing between that which emanated from the commercial interest and that from the Department (purportedly the national interest). Statistical information has been falsified to favour the Road Lobby interest, has been fed to Ministers, and has thereafter been quoted by Road Lobby spokesmen as 'Ministerial statements'; and should individual ministers at any time appear to assume some independence of the Department/Road Lobby machine as Mr Hamer states: 'the Lobby may then proceed to the second stage, political pressure, in which it threatens to upset the political power base of the Minister, or even of the Government as a whole, through influencing M.P.s . . .' And to further this end: 'The A.A. and the R.A.C. alone estimate that they are in contact with 100 M.P.s within a single Parliamentary session. For a major debate the A.A. will brief about 75 M.P.s' (op. cit.).

And, of course, there is the use of Public Relations Officers. Many of the multifarious Road Lobby members have their own, but undoubtedly the most powerful is the lobbyist of the British Road Federation itself, Lt. Commander Christopher Powell, who has held the post since the war. How is it, Sir, you may well ask, that these lobbyists, in close co-operation with the Department Civil Servants, are able so successfully to destroy the representative and legislative function of Parliament in transport matters? Chiefly, Sir, as I have already indicated, by the control of information. With the closest co-operation of their friends in the Department, these lobbyists, clearly with very substantial budgets, are able to decide precisely what information M.P.s shall receive (or shall not receive, in the matter of Green and White Papers and answers to questions), to distribute information to Members apparently bearing the imprimatur of the Department, and to give outright misinformation where necessary should Members persist in demanding information which might prove an 'administrative difficulty' for the Lobby (such as when Mr Mulley, Transport Minister as he then was totally misinformed the House on May 7th last year on the all-important matter of the manner and extent of Government grants to the British Waterways Board). (The above, Sir, can all be substantiated from any close scrutiny of Hansard over the last ten years, as my submissions to inspectors at previous inquiries have gone to show.)

But to return to Lt. Commander Powell. My clients require to know, Sir, and request the publicising of this information by the Minister (for the country requires to know, so widespread is public disquiet on these matters) what precisely are the limits to this retired naval officer's influence? Apparently, since the 1940s he has been the Secretary to the unofficial All-Party Parliamentary and Scientific Committee. As this Committee includes transport matters within its purview, my clients request to know what financial backing is available to him and from whence, to further the Lobby's ends through the Committee's activities. What is his role, they require to know, within the All-Party Roads Study Group? Is he and the Lobby he represents responsible for the lavish hospitality, the cocktails, lunches and dinners afforded to M.P.s, Departmental Civil Servants and Road Lobby representatives which is such a characteristic of the meetings of this Group?

Does this lavish hospitality end there? Is its objective the securing of votes in vital debates?

How wide does his influence extend? Mr Frank Allaun M.P., writing in the *New Statesman* of November 1973 records the following curious event:

> I can remember one occasion some years ago, when the Lancashire group of M.P.s, of which I was then Chairman, received an invitation to dinner to discuss the future of the County's roads, ostensibly from the Lancashire County Council. However . . . while the County Council had doubtless concurred, the invitation came from a retired naval officer who was the chief Public Relations Officer of the British Road Federation, who were the real host and footing the bill.

Lieut. Col. Gerald Haythornthwaite, Technical Adviser to the Peak District Branch of the C.P.R.E., tells a similar story regarding the visit of Yorkshire M.P.s to look into the proposal for a Sheffield to Manchester motorway. Ostensibly it was a Departmentally organised occasion. It turned out to be organised by the British Road Federation, who not merely organised the luxurious limousines and lavish hospitality, but also the carefully selected route, which, significantly enough, was mainly along the already constructed M62.

But, of course, it cannot end here, can it, Sir? So many decisions on transport are not the responsibility of Central Government and the House of Commons, but Local Government and those local councillors comprising the Transport and Highway Committees. May we take an example of how things have been managed, apparently, in London?

During the late 1960s Mr J. M. Thomson produced a well-researched, statistically detailed attack on the proposals for the Inner London Motorway Box, entitled *Motorways in London*. A letter, apparently coming from G.L.C. files, dated 10 November 1969 includes the following:

> Dear Phillipson
>
> As arranged on the telephone, I enclose some comments on *Motorways in London*. They are only first thoughts and initial technical reactions. You appreciate they are not intended for publication and they do not represent the Council's considered view on the book. However, I hope the comments on this book will serve your immediate purpose . . .

The letter followed with ten pages of detailed criticism of *Motorways in London*, all vital to the British Road Federation in its campaign to discredit the opponents to the construction of the motorway box.

When it is borne in mind that the Phillipson to whom the letter was written was and is the Director of the British Road Federation and that the person who wrote and signed the letter was a Public Servant, we see here another example of the Road Lobby 'needing the help of the Department' (in this case a department of the G.L.C.) and having no difficulty in obtaining it.

Sir, at this point, I am required to say that under these circumstances there can be no doubt that what we have in these road programmes is nothing but the contrived manipulation by both central and local government Civil Servants, acting together with agents of the Road Lobby, of the whole democratic process of decision-making in transport matters in the exclusive interest of that commercial lobby.

But, Sir, in saying so, it is not my clients' intention in any way to attribute venality or any motives of personal gain to any public servants or any other persons involved in this national outrage; of this, so far as I know, there is no evidence. But while disclaiming any such motives, what my clients wish me most powerfully to put to you, Sir, is that, though personal venality may play no part in their actions, a besotted adherence to an outmoded form of transport and a wholly brutalised attitude towards human communities or anything that stands in the way of their highways programmes, together with a calculated denial of adequate investment programmes to any alternative transport modes that might threaten the Road Lobby position, has had the most disastrous effect upon this nation, its social fabric and its future economic prospects – and that in this sense the actions of those Civil Servants in the Department are wholly culpable.

But again, Sir, the matter cannot be said to end here. This matter of Departmental functional corruption is now so serious that my clients wish me to submit to you that the nation at the moment has no guarantee whatever that the elaborate national consultation procedure initiated by the Secretary of State in the Orange Paper will be nothing more than a motorway inquiry writ large. And by that I mean that in just the same way that you, were you to allow this Inquiry to proceed, would no doubt take down in painstaking detail every submission against the proposal, only to send your report to the same Highways Directorate of the Department who, with the assistance of the Road Lobby, have proposed the scheme in the first place, and thereby ensuring that all those objections are ignored in the interest of what the Department/Road Lobby machine is pleased to call 'policy'– so, in just the same way, all those meticulously detailed submissions in response to the Orange Paper from the C.P.R.E., the British Waterways Board, Transport on Water Committee, British Rail, the National Federation of Womens' Institutes, the Conservation Society, to name but a few, will be received by the same Secretary of State (or whatever the new incumbent's title now is) whose immediate predecessor has just obliged the roads interest by signing the Line Orders for the greatest single commitment to future motorway construction for a decade (the M40/M42 Orders), and again, in the name of 'policy', treated with the same contempt. In other words, just as Line Orders are signed, committing huge and irretrievable resources to motorway construction, thereby undermining the economic viability of major segments of the rail and waterways systems, despite the admissions of the Orange Paper that we still lack even the *objectives* of a national transport policy, so, while the close link between the Department and the Road Lobby remains unbroken, we shall see no consideration whatever given to the national response to the Orange Paper.

Indeed, Sir, the nation has already been served clear notice that this will, in fact, be the case. In noting that 'The Department has often been criticised for lack of co-ordination, notably between road and rail, and has been accused of being unduly influenced by the "roads lobby" ', *The Times* of the 21st June of this year [1976] went on to announce that the Department had set up a Steering Group under Mr Peter Baldwin, to co-ordinate the various transport and planning interests. Two things, however, suggest powerfully that nothing is to change. The first is the membership of the Group, who with only one exception are known committed road supporters. And the second is the bland admission by its appointed Chairman, Mr Baldwin, the Permanent Secretary, that in his view apparently there really is no need for the Group to exist or be steered anywhere, when he said, as reported in *The Times*: 'People believe that because this Department has a road programme to carry through, it is pro-road and anti-rail. It is not. We are pro-transport, in the best attainable environment.'

I suppose, Sir, that had I been asked to put together 32 words to make it quite clear to the country at large that the objective of the Group was to ensure that there was to be no change, I could not, had I thought about it for a week, have chosen more appropriate words. While again imputing no ignoble motive to Mr Baldwin, whose words probably indicate no more than an obstinate determination to protect the *status quo*, I put it to you, Sir, that were I Mr Phillipson or any of his colleagues in the Lobby, I should feel that I could rest more easily o'nights having read them.

And so, Sir, to summarise: there comes a point where lack of confidence and trust has become so wide, so deep with, to use the words of the Bishop of Winchester, so many decisions 'made behind closed doors', that untold damage is done to that complex, delicate fabric that constitutes the modern democratic state. And that, Sir, in my submission to you, is the point that we have now arrived at in the matter of transport planning in this country.

The function of planning has been so corrupted; the link between the Department and the Road Lobby is so complete; the denial to Parliament of access to impartial information and the destruction of its powers thereby is so total that a very large segment of this nation has now lost all faith. They have no faith in your office or its supposed independence of the Department; they have no faith in the Secretary of State or the now Secretary for Transport, whomsoever occupies that position, seeing him as no more than a mouthpiece for those Highways Civil Servants whose sole object is, to quote again from the *Architects' Journal* to ensure that when he pronounces on transport 'he speaks with the mouth of the road engineer'; they have no faith in their Parliamentary or local government representatives, whom they see as no more than helpless, manipulated puppets of the Department/Road Lobby machine that has, in the matter of both central and local transport planning, achieved no less than the destruction of the representative function of British democracy.

And so, Sir, my clients have but one recourse. All that they can do now is to stand up here before you and say: We have had enough; we are taking no more. And if it is your intention here today, Sir, to ignore my appeal to you, and to use brute force in the interests of that Department/Road Lobby machine, then, Sir, I am required to tell you on their behalf, they will not resist. They are aware, Sir, of the extent to which they are threatenened; they recall that at the Aire Valley Securicor personnel were authorised to use force against the objectors, but in the matter of the use of the police, Sir, they know them to be misused and they have no quarrel with them; they will offer no resistance.

But, Sir, know you this on my clients' behalf: they will be back. They will be back in increasing numbers tomorrow and the next day and the next. They and their fellow men and women in Yorkshire, in Winchester, in Sussex, in Essex, in Lancashire, in Cheshire, in Oxfordshire, and increasingly in every shire in the country, they will be there, interposing their bodies and their plain courage between you, Sir, and your like (should you insist upon lending your support to this monumental evil) and the injustice and cruelty that would be perpetrated.

But, Sir, those are only my words. Others express these matters differently, but the message is the same. The *Solicitors' Journal* for August of this year includes some comments by a Mr Rose. This is something of what he had to say:

> In the context of the monstrously inflated power of central and local government to enrich individuals, is it credible that all the tycoons and traders involved in road transport and road construction, in the supply of materials and the production of motor vehicles, sit neutrally aside and take no steps to make friends and exert influence among the civil servants and the politicians who control this Aladdin's cave stuffed with multi-million pound projects?

And he goes on to say:

> The politicians and Civil Service mandarins should start re-thinking the effects of courses which are undermining respect for the rule of law in habitually law-abiding people . . . The consequences of the Ministry's motorway policy on the lives of millions of people are far more momentous than any but a very few of the matters judicially decided in the courts of justice. The sense of frustration and injustice so generated is calculated to undermine the whole habit of obedience to constituted authority.

Now, Sir, I wonder if at this moment I could save us all time and anticipate something of what Counsel for the Department may have in mind to say when he gets on his feet. Perhaps he will say, Sir, as so many of his often very eminent predecessors have done in the past, that these matters are not of your concern; that you have been appointed by the Minister, under the various provisions of the Acts, to conduct this Inquiry, and that the limit of your duties regarding the matters that I have raised before you is to give an undertaking that you will refer them all to the Minister himself. I shall deal with that matter in a moment, Sir. But perhaps he will also say that the matters that I have raised are not your concern, nor even that of the Minister himself, but that they are matters for the Courts of Justice to decide, and it is there that Mr Tyme's remedy must lie. Sir, in a sentence or two, may I dispose of this matter.

For over three years I have been engaged in these highly contentious and litigious matters on behalf of the National Conservation Society, many other organisations and many hundreds of individual objectors. Well over a thousand pounds has been spent in discovering whether or not this attractive sounding remedy is open to us. To quote from a legal summary on a related matter regarding the Secretary of State's refusal to comply with a vital clause of the 1959 Highways Act (a matter eminently more appropriate to the Courts than the matters before you now, Sir, and upon which we have the support of several leading counsel):

> . . . you can take no proceedings in the Courts with any hope of success. . . . English law relating to administrative matters is in a mess and needs thorough reform and rationalisation.

But, Sir, in my final words, I address you with all the gravity at my command. On the opening day of the inquiry into a motorway-standard by-pass in Skipton in Yorkshire, the Inspector, a Mr Rolph had many of these matters raised before him. Apparently, Sir, he seemed to regard them with what almost might be termed levity. 'I have been appointed to open this inquiry, and by jove, I am going to do it.' I confess, Sir, not to know what informs such an attitude of mind, and cannot find words to express the gravity of the state of this nation, when inspectors appointed by the Secretary of State to make recommendations on matters of such enormous national consequence show by their attitude and their words such total irresponsibility. I have it, Sir, that that 'independent' inspector did not give the matters so much as five seconds of his time and consideration.

And so, Sir, it is with that recent background in my mind that I now close my address to you. Sir, with every atom of my being, I pray that you might (as did Mr Harcourt, who resigned from the Road Inquiry Inspectorate in disgust at what he had been asked to preside over at the Inquiry into the Yeovil Inner Ring Road) rise to this unique opportunity to strike a blow for natural justice and – no less – for the preservation of this democratic and Parliamentary system by which we are governed, and thus help to reverse that process of national alienation referred to by Mr Rose.

I beg of you, on behalf not only of all my clients here, but on behalf of those thousand upon thousand people across the face of this land who see the process of transport planning and those inquiries into their proposals as no less than a fundamental assault upon the hard-won rights and freedoms under the law by which we are still privileged to exist; an attack by a combination of bureaucratic power and immense financial interests upon our precious democratic heritage – I beg of you to adjourn these proceedings at once and make your personal recommendation on my clients' behalf to some independent judicial authority – the Lord Chancellor, perhaps – so that all these similar proceedings across the face of the country can be terminated at once to await an investigation at the highest level into the grave matters that I have raised before you.

* * *

The major upset on the opening day with an adjournment in the first sixty seconds, and the sit-down demonstration on Day 3 were led and organised by me. An article in the *Sunday Telegraph* for 19

September under the headline 'Tyme machine moves on against motorways' gives the flavour of the events:

The tiny and turbulent force of John Tyme, a 50-year-old veteran anti-motorway campaigner, moved into a public inquiry into a road extension at Hornsey last week, and the inquiry finally broke up after he told the inspector he regarded him as 'a bulldozer' and led a sit-in at his feet.

The sit-in came at the end of a carefully orchestrated three days of verbal mayhem and disruption by The Tyme Machine which, at one stage, had the inquiry inspector, Mr James Vernon, reading the riot act to the audience.

'This is not a forum for demonstrations,' he said. 'The police will be called. Like Notting Hill this hall is not going to be a no-go area.'

Despite the inspector's stern warning of Friday morning Mr Tyme – who had steered a stormy course through such inquiries as the M42 (Bromsgrove), M20 (Kent), M16 (Epping), M40 (Kenilworth), the Aire Valley by-pass and the M3 (Winchester) – denounced the inspector and the proceedings and, after quietly packing up his case, sat down on the floor.

Some 100 others joined in the sit-in and, to the rousing refrains of 'We Shall Overcome' the police escorted them one by one out of the hall.

Mr Tyme is a slight, silver-haired man who enjoys studying moths on his holidays. Over the three days at Hornsey, one gained an insight into the personality and methods which have made him a sort of Che Guevara of the North fighting the Batista of the growing motorway network.

To date he has pitched into 17 motorway inquiries, using a mixture of legal erudition and downright rudeness.

In the inquiry itself at Hornsey his methods were as varied as ice and flame and he fought it out with the inspector in much the same way as a matador attacks an extremely clever bull, amply aided by some local picadors. The proceedings broke up and are likely now to continue to be broken up until the plan is abandoned.

The local anti-Archway motorway group had been planning their opposition to the road all this summer and Mr Tyme was enlisted to spearhead the fight against a mile of dual carriageway which involves removing 120 houses and shops at a total cost of £11 million.

Immediately Mr Tyme moved in, battle plans were drawn up at private meetings. It was agreed among the protestors that on Day One of the inquiry Mr Tyme should stand up right away and argue that the inquiry was illegal.

The inspector refused to let him speak and, in the ensuing uproar, adjourned the meeting. Then Mr Tyme, the diplomatist, arranged a backstage meeting with the inspector who said that providing the meeting was orderly, he would be prepared to listen to Mr Tyme's and other objections.

At lunch-time on Day One Mr Tyme and his supporters retired to a local cafe to discuss tactics . . .

In the afternoon he was finally allowed to make his speech and on Day Two the inspector let the other objectors speak too.

After contenting himself with some occasional heckling Mr Tyme exploded, accusing the inspector of being unable to perform his duties impartially because of his former association with the Department of the Environment. He angrily called on the inspector to resign.

On Thursday evening Mr Tyme and his lieutenants decided that, as their numerical support in the Guildhall audience was dwindling, they would try another more direct tactic on Day Three. They agreed they should try to get a statement from Mr Vernon that either he resign or declare the inquiry illegal.

When Mr Vernon refused to do this Mr Tyme acted – after first apologising for giving the inspector a headache the day before. 'I feel a despair wavering on anger,' he said. 'I regard you as no less than a bulldozer and sit down in front of you.'

After the sit-in, Mr Tyme – who is much politer to the Press than to officials from the D. of E. – declared himself satisfied with progress and went off to be interviewed by the B.B.C. and on to Canterbury and another motorway battle.

But my immediate departure thereafter, first to Canterbury and then to Chichester meant that responsibility for events now devolved upon the many very able people in the area: Dr Peter Levin, Dr John Adams, George Stern, Terry Rand, Bill Tyler, Sally Vernon and others. It was they, together with the support of hundreds of objectors, most of them women, who forced the inquiry to a standstill after fifteen days. It was reported at the time that the inspector who had threatened to take the inquiry into 'private session' was ill, but the long adjournment of almost seven months and the appointment of another inspector suggests that other reasons may have been at work.

* * *

The reopening of the inquiry on 19 April 1977 under Mr Rolph recommenced the battle all over again. Once again my role was merely an initiatory one. By now the dominant themes requiring to be emphasised had become the decision-making machinery within the Department of Transport and the lack of Parliamentary accountability of the Secretary of State.

Accordingly, the submission began by asking the question: what was the position of the Secretary of State in the matter of the present inquiry, and answering it by saying that he was 'ultra vires', beyond his powers. In saying this, I said that my clients were not denying his powers to convene the inquiry under the Act, but that they *denied his right to exercise them.*

The reasons for this were two-fold. First that, as it could readily be shown that Parliament has had no control whatever over the road programme (which constituted a denial of that fundamental principle of the government of this country that Parliament should exercise control over the executive, and in most particular, over expenditure), it followed that the Secretary of State could not now draw and exercise powers from Parliament as set down in the relevant Act. He could not, in other words, have it both ways. Either he granted Parliament control over the motorway/trunk road programme (by setting before it at the appropriate time White Papers relating to it, and by granting it full powers to debate and authorise – or not authorise as the case may be – the financial provisions for it); in which case, it was right and proper for him thereafter to exercise his powers under a Parliamentary statute relating to that system; *or* he denied Parliament such control by denying it White Papers and means of debating and approving the financial provisions; in which case, it became a constitutional abuse to exercise such powers. Or what objectors were saying in blunter terms was that if the Secretary of State denied Parliament its rightful and fundamental role, then he should proceed with his road and motorway proposals with equal candour. He should not come quoting Acts and Statutes at inquiries through the mouths of inspectors, for, having denied Parliament the *general*, he could not set up inquiries, flourishing from Parliament the *particular*. And that if he did, he would be rejected.

And the second reason was that, while acknowledging the title of the Right Honourable William Rodgers, namely that of Secretary of State for Transport, objectors believed that in the execution of his duties persuant to that office, he was circumscribed in such a way that he could not carry out such duties to Parliament and the country in terms of *transport*, but only in terms of *highways*. And the reason for this, they believed, lay in the structure of his department and the decision-making machinery therein.

I then went on to say that this second argument was perfectly exemplified by the situation relating to the inquiry and the Line Order Inquiry which preceded it. The subject of those inquiries was a basic problem, namely the provision of facilities for the movement of people and goods throughout north London, clearly a *transport* issue, involving the consideration of all modes, rail, road and waterway. The manner in which the Secretary of State dealt with the subject, however, was to hold an inquiry into a proposal which excluded consideration of all other modes and proposed but one solution, namely the provision of more road space. And the explanation of his failure to deal with this problem, as in countless other cases throughout the country, as a *transport* issue could only be explained by the manner in which he accepted his executive authority, allowing his transport duties to be replaced by his duties as the national highways authority, and allowing the decision-making process within his Department to reflect that acceptance.

I went on to say that within this context my clients asked themselves this question: How, when the Secretary of State's own Permanent Secretary was Chairman of the British National Committee of the Permanent International Association of Road Congresses, and when no such executive sat on any similar committee relating to other and competing modes – how under these circumstances could the Secretary of State exert that parity of interest and concern between competing modes which, by the nature of his office, it was his duty to ensure? And they answered the question by saying that he could not, which constituted the full and entire reason why the provision of facilities for the movement of people and goods in north London was conceived, not as a transport problem, but only as a highways problem.

I continued by pointing out that the Secretary of State had conveniently made the point himself when, in a letter to the Right Honourable Frank Judd M.P. (9 February 1977), he had written:

> It [membership of the PIARC] again stems from the Department's executive responsibilities. That there is no rail equivalent only reflects the different organisation and problems of the railways.

This, I said, explained the situation precisely: while the Secretary of State's executive responsibilities were bound up with the PIARC, no such similar responsibilities could be made to apply to rail and waterways, which *in no similar sense whatever* were the Secretary of State's executive concern, and that under such circumstances, parity of interest and concern was impossible.

At this point, I sought to anticipate what Counsel for the Department might say in his rebuttal, and said that, in the matter of Parliamentary control, it would avail us nothing to be told of Treasury White Papers and Central Government Expenditure Statements. These, with their arbitrary allocation of funds to one thing or another, granted to Parliament no opportunity of detailed examination or debate and were no more than a symptom of the disease. It would not help either to be told that the inquiry was not the proper place to raise such matters, that they should be raised in Parliament, that the Secretary of State's responsibility to Parliament was not the Inspector's concern, that in so far as he was required to do, the Secretary of State had complied with his requirements under the Act or that the Courts should be applied to, etc. etc. – for we were at what could only be described as a constitutional watershed.

I acknowledged that, of course objectors accepted that Parliament was the only place where the issues could be resolved. But equally, they accepted the need for the most exceptional action in order that Parliament should become that place, which currently it clearly was not. I quoted Edward du Cann M.P.:

> . . . the chief protection of the rights of the citizen, Parliamentary control of the executive, is decreasingly effective. M.P.s cannot even control public expenditure.

and said that if those M.P.s who agreed with Mr du Cann (and they were many on both sides of the House) were right, and if, as was the case, they had been saying much the same thing for decades, and if, as could be shown, far from getting better, the situation was getting steadily worse – then my clients acknowledged that Parliament was in need of some assistance. They accepted that many M.P.s needed a reminder of the tenuous hold that was left to them of their rights and powers, *and that, by their stand, my clients undertook to provide it.*

This submission, when rejected by the inspector ('I will convey its substance to the Secretary of State'), was followed by an 'occupation' and major police intervention. In the struggle over the ensuing days a major concession was gained in the form of transcripts, the free copies of which have since proved quite invaluable to objectors seeking to follow events. When, however, a paper was circulated in the inquiry questioning the inspector's independence, he announced that the proceedings would continue in a room closed to the public which became known as the 'bunker'. There, with a handful of 'representative' objectors present, it pursued its chaotic course for a week. The *Hampstead and Highgate Express* editorial for 27 May records:

> . . . the sight of three screens of police and commissionaires guarding an inspector and Ministry officials while the latter supposedly conduct a public inquiry through inefficient loudspeakers is on Orwellian nightmare.

Once again it became a public inquiry for a while, but when the inspector attempted to terminate submissions on scope prematurely, further disturbances occurred involving the police and the extraordinary accusation by Counsel for the Department that the objectors were supporters of the Berlin Wall. This led to a restoration of the Orwellian nightmare. This time no objectors at all were allowed in and an attempted protest led to violent assault upon objectors (one of them a G.L.C. councillor) by a Ministry official. An adjournment followed until June, when it was reopened, again in private session. In the following month, July 1977, there occurred Haringey Borough Council's decision to object to the proposals, which led to a substantial adjournment and to highly significant developments.

* * *

At the reopening, in public again on 3 October, I was escorted out by the police, having been refused permission to read a submission which began by saying that I had been asked to address him (the inspector) at that juncture because of the gravity of the situation, for, some weeks previously, physical violence, assault upon the person, had taken place in his inquiry. Objectors, I reminded him, in no way violent themselves, but frustrated almost to the point of despair by the course the inquiry had taken, had been subjected to physical violence and, it was alleged, in the case of a G.L.C. councillor, punched in the mouth.

I said that nothing could better illustrate the accuracy of the warning that I had given on the opening day of the inquiry over a year ago, when I had quoted words from the *Solicitors' Journal* for August of that year referring to courses of government action which were 'undermining respect for the rule of law in habitually law-abiding people' and 'undermining the whole habit of obedience to constituted authority'. The warning, I said, had gone unheeded. I made the point that I was sometimes led to think that I was addressing the empty air when addressing inspectors at road inquiries, as they seemed to deny any personal responsibility for what happened or what should happen in their inquiries – on the assumption, apparently, that they could shift all responsibility onto the Secretary of State. I sought to impress upon him that I was addressing him in that instance not merely as Inspector, but also as a fellow citizen of this democracy of ours and one who could not look to such developments as quoted in the *Solicitors' Journal* with either favour or confidence.

I said that I wished to raise two matters.

I first wished to show how the inquiry and particularly his appointment to preside over it was a denial of the second of the two fundamental pillars of natural justice, enshrined in the well-known words *nemo judex in causa sua*: let none be judge in his own case. In order to do this I drew from my current experience in taking a major case at the inquiry into the proposal to build a thermal oxide reprocessing plant at the nuclear research centre at Windscale. And I asked the question: What possible credence could have been given to that inquiry, what possible faith could all those major objectors have had in its deliberations had the inspector, Mr Justice Parker, been known to have spent the best part of his professional life supervising the building of just such plants, and had over recent years presided over a dozen or so inquiries into proposals to build them, invariably recommending in favour of them all? The notion, I said, with respect, was so preposterous that it did not bear a moment's consideration.

Yet that, I submitted, was what my clients were faced with at the Archway inquiry in his appointment as Inspector. I made the point that in no way did such a statement call into question his personal integrity as indeed it would not in the case of Mr Justice Parker. It merely meant no more – but no less – than that, nurtured, cocooned within the specialised world of a given technology, be that nuclear engineering or road building, neither of them would be able, despite the most sincere efforts, to establish that impartiality, that total independence of mind which was the quintessential part of the responsibility that their appointments laid upon them – nor, significantly more important, would they be able to convince the concerned public that they had been able to do so.

I then raised a closely related issue, quoting the *Nicholson* v. *Secretary of State for Energy* case in which Sir Douglas Frank had clearly enunciated the interpretation that in a local inquiry people are now expected ' to take an active, intelligent and informed part in the decision-making process', and said that if it had been impossible before that interpretation to see his appointment as anything but a denial of natural justice, now, it could not be seen as anything but intolerable and absurd. For, I asked, with respect, how could he, within the narrow terms of his experience, within that linear progress that he had made from one road decision to another, how could he *conceivably* be expected to preside impartially over an inquiry with such a radical interpretation of its function? And I answered by saying that he could not, and that nothing in this world would convince my clients or those tens of thousands of concerned persons throughout the country to the contrary.

I then proceeded to state the requirements, which were two: *First* that he should withdraw from the inquiry; and *Second* that the inquiry be terminated under whatever statutory powers necessary and that a fresh inquiry be convened as a *transport and planning* inquiry with detailed considerations within its terms of reference as to how modes of transport other than roads could relieve the area of the atrocious and intolerable conditions that road transport had imposed upon it.

I then went on to say that if these conditions were not fulfilled, my clients would, with the utmost rigour and endless determination, use every means at their disposal: *First* to understand the scheme; *Second* to discover why it was necessary within the whole complex of transport and environmental planning; and *Third* and most importantly, to discover what conceivable transport policy now having Parliamentary and government approval could be said to justify and support it. And that, having established all that, they would then proceed to object to the proposal, calling upon all the increasing rights and powers which came to be at their disposal.

I concluded by saying that *finally*, notwithstanding all of the above, because of their total lack of faith in him, the inspector, and their absolute conviction that neither their own objections, nor those of Haringey Council (no matter how valid) would count for a fig within that functionally corrupt Department of State that had set up the inquiry in the first place – because of this, should the Secretary of

State then authorise the scheme to go ahead, they would regard his powers as having no legitimacy, no legality, and his action as an open invitation to take any measures necessary in defence of their homes and their environment, and thus to assert their rights under law and natural justice that would have been denied them. And that, I concluded, would bring us grimly back to those extracts quoted from the *Solicitors' Journal*.

The submission, or most of it, was read in my absence by Diane Rudin, but of course it was ignored. This was of little consequence, however, as the inquiry now took a new form, occasioned by Haringey Council's decision to object to the proposals. Two points must be made regarding this decision. Undoubtedly it was due to the determination and endless resistance of the objectors, so ably led by Peter Levin, John Adams, George Stern, Bill Tyler, Diane Astrop, Tony Richardson, David Gleave, Colin Bex and others. The initial check at the opening, the agreement to discuss the widening scheme itself and the very favourable media publicity that the objectors had gained gave the local councillors time to re-examine the implications of the scheme and make their all-important decision not merely to object, but to employ for the presentation of that objection the expertise of Metra Consultants Ltd. The second point follows on from this, for Haringey Counsel's skilful cross-examination, based upon the consultants' detailed and meticulous work, began to reveal what objectors had suspected all along – important unacknowledged changes and falsifications in the Department's case. As George Stern puts it in an article in *Transport Retort*, December 1977: 'They knew the changes well enough, so the only conclusion to be drawn is that they hoped the inspector would bulldoze their phoney figures through . . .'

Furthermore, it became clear as the remorseless cross-examination continued that *the purpose of traffic analysis and prediction is not to arrive at any real understanding of the problem and possible solutions, but simply to justify a road-building scheme already decided upon.*

At this point also the inspector ordered the publication of the 1971 Consultants' Report on the widening of the A1 through Hampstead Garden suburb. This highly important report, crucial to the future of so many Hampstead residents and highly relevant to the Archway inquiry, has remained secret for the past six years, despite every effort of the Hampstead Residents' Association to have it published. The Secretary of State, however, refused once again to release it, which led the *Hampstead and Highgate Express* for 25 November to make the following comment under the heading 'Star Chamber':

What's more, he brushed aside what was an ORDER from his own inspector and referred to it as a REQUEST. No wonder the objectors now declare that this must bring the Archway Road inquiry farce to an end.

Objectors have been promised details about traffic forecasts, cost-benefit analysis and environmental impact, but so far these have not been produced by the Department and the inspector has indicated that if the information is not forthcoming he will adjourn once again; in any case, should the information become available objectors will clearly need some months' adjournment in order to prepare a case based upon it.

Back, however, to Haringey's cross-examination; at the time of writing the inquiry is once again adjourned and *sine die*. The relevant and highly significant words of the inspector in declaring the adjournment are as follows:

*From the evidence given it is clear that there have been changes of sufficient magnitude that would justify the recomputation of the predictions . . . I consider that this step would be best, both in the interest of the Secretary of State and the objectors. I am sure that Government would not wish to go ahead with such an expensive scheme unless it is clear that future traffic usage will justify it . . .**

The 'changes of sufficient magnitude' are the falsifications and errors. One can only wonder how many 'expensive schemes' Government *has* gone ahead with for the want of adequate scrutiny.

* This inquiry and the proposed schemes and orders have since been abandoned.

6

The Ripley Inquiry: Blank Cheque for the M25

Therefore is judgement far from us, neither doth justice overtake us: we wait for light, but behold obscurity; for brightness but we walk in darkness.

Isaiah 59:9

The inquiry into the Wisley Section of the M25, that six-lane highway proposed to encircle London in order to receive the fourteen six-lane highways proposed to descend upon London, opened at Ripley on 15 November 1977. The inspector was Rear Admiral Nixon.

At the inquiry I acted not for local objectors, but for the Conservation Society, and the issue there raised, which reveals one more facet of the whole corrupt system, is explained in the submission, the contents of which are as follows:

I began by stating that at two major road inquiries, with the early stages of which I had been closely involved, it had subsequently come to light that evidence from the Department in support of the schemes had been shown to be invalid and, what is more, known to be invalid. I mentioned first that at the M3 inquiry at Winchester it had been discovered by Mr Gordon Macpherson, an emminent investment analyst who happened to possess the necessary expertise to work according to the Department's book (a process involving a massive analysis and accountancy exercise) that incorrect figures had been used in the cost-benefit analysis (C.O.B.A.) calculation, and that when the correct figures had been substituted, *the C.O.B.A. value went negative*. I then produced evidence of the changed figures at the Archway inquiry and quoted the inspector's words announcing the consequent adjournment.

I pointed out that this raised a vital question for my clients and for other objectors at the inquiry. For if the Department and its representatives in the Road Construction Unit could be shown within the space of six months to be using false figures to justify two major road proposals, then what guarantee had the nation at large and in particular those persons present whose properties were threatened by the Wisley proposals, that similar practices were not employed there?

I pointed out that the nationwide distrust in the Department of Transport could not help and that the lack of any independent audit system within the Road Construction Units (a situation that no

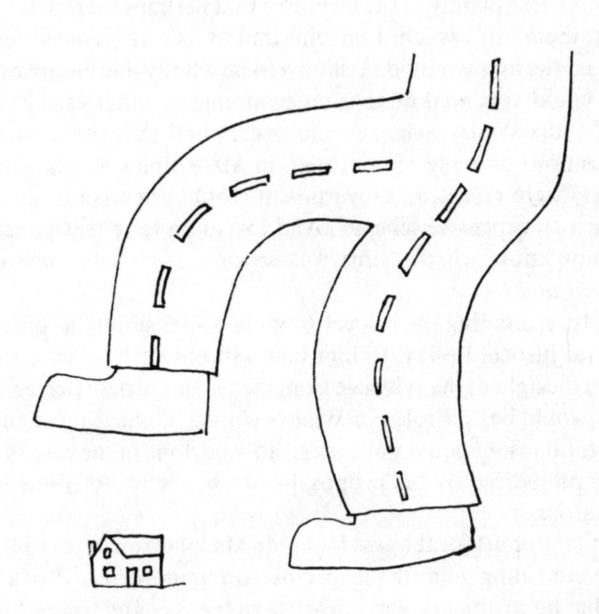

private firm would tolerate for a moment), the denial to Parliament of access to impartial information by which alone it could exercise adequate control of the Department's activities, and the manifest incapacity of inspectors at road inquiries to unravel the complexities within which those major errors or deceptions were found – all had led to a situation in which a large segment of this nation had lost all faith.

It was in those circumstances, I said, that what I was respectfully suggesting was that the national interest required *an independent audit in detail* of both the traffic forecasting and cost-benefit statistics upon which the proposals before him were based *and that that examination should take place prior to the inquiry's hearings.*

I drew from quotations regarding the 'audi alteram partem' rule to ask how objectors at that inquiry could know the case against them and thus be heard when, because of the chance absence of a Mr Macpherson or an objecting local authority, they could not be sure of the facts of and thus the true nature of the case against them. At this point I put very plainly to the inspector that perhaps there *was* no case against them, by which I meant that if, as at Winchester and Archway, the figures could be shown to be wholly inaccurate or false, then it could very well be that no economic or other case could be made for the Wisley Schemes and orders and that section (or any other section) of the M25, and that, in Mr Rolph's words, once the true facts were revealed, 'Government would not wish to go ahead with such an expensive scheme'. And I went on to reiterate that they could not know whether this was so or not *without such an independent audit.*

Finally, reminding the inspector of his responsibility as custodian of natural justice, I put it to him that without such an independent audit, in the light of the evidence from the two inquiries referred to, his inquiry would be, in Professor Wade's words, 'no hearing at all', and that natural justice could not conceivably be done in the case of those persons prejudiced by the proposals, nor be seen to be done by the nation at large.

Despite support for the case from Mr Macpherson in person, in his subsequent ruling what the inspector said amounted first of all to a claim that his arithmetic was adequate to the task and furthermore to a promise that if he found that the figures warranted it, he would then order whatever audit appeared to be necessary.

Some weeks afterwards, I received a copy of a letter sent by Mr Macpherson to the inspector, and I have received permission to quote from it:

> I have been informed, as a result of a telephone call to your programming officer, that you have decided not to adjourn the Inquiry for submission of the traffic data and cost-benefit analysis to independent audit on the grounds that until you have heard the evidence on these items you are not in a position to know whether or not such an independent audit might be required . . .

In my submission, it is not reasonable to expect that such a complex matter can reasonably be investigated in a satisfactory manner within the current public inquiry system. It was for this reason that I supported Mr Tyme's request that such matters should be subjected to independent professional investigation and audit. Until such an investigation and audit takes place, you and the public are required to accept the Department's computer-derived figure on trust, even though we know it to be subject to unknown degrees of error. In my opinion a potentially dangerous policy is being followed if future inquiries into road proposals continue to be required to accept such computer-prepared figures on trust, and the danger arises from two features that are inherent in the use of computers.

Computers are massive mindless calculating machines that suffer from two important failings: they are vulnerable to operator error, and they are feared by the public. Both of these shortcomings arise from inadequate understanding, but are no less real for that. Operator error becomes more difficult to detect as programs become more complex, whereas public fear grows with the size of the monster which appears to be under unknown control. I raise this subject, not because I am opposed to the use of computers for traffic models and economic appraisal, but because I believe that road schemes are so complex that electronic data processing is absolutely essential in their preparation. The requirement and need is for a system of checks and safeguards to permit the use of computers in such a way that will control the extent of error and will provide grounds for greater public confidence in their output.

This is not a question of whether one is for or against motorways in principle, but it is concern for the correct application of government investment funds and for the maintenance of trust and confidence by the public in our system and processes of government decision-making.

Once trust in government breaks down, fear too easily can enter to fill the vacuum. The first victims of fear tend to be law and order, as has been seen so often in the course of human history. My limited experience of the M3 proposals and even more limited experience of the M25 proposals, give me reasons for believing that public trust is wearing thin, and this is particularly the case in the area of data produced by computers . . .

Evidence is now coming to hand of other C.O.B.A. scrutinies revealing errors and inconsistencies similar to that at Winchester. The Leitch Committee, whose Report at the time of writing is due to be published in a matter of days, is known from leaks reported in *The Times, New Scientist* and *Sunday Times* to have found major inadequacies in the Department's forecasting methods.

* * *

Had the issue of the independent audit been raised at Winchester or Archway, it would have been fought for and won. But at Ripley the vital equation was unbalanced. People had come from as far afield as Chichester, Winchester, Archway, Ipswich and Hornchurch, but they were a mere handful. The inquiry, therefore, went ahead. The inspector will write his report to his masters who will duly have the Orders signed and contracts let at the earliest moment. The equation cannot always be balanced, and unless something fundamental is done this huge and irretrievable investment of resources in unproven schemes will go on indefinitely and thus make inevitable that transportation collapse alluded to in Chapter 1. That something must be done is clear; that something can be done and what and how is the subject of the following chapter.

7
Corruption of Government

I am willing to hear your grievances, as my predecessors have been, but I must let you know that I will not allow any of my servants to be questioned by you . . . Your business is to hasten and grant me the supplies I ask or else it will be the worse for yourselves.

To the Commons at Westminister, from Charles,
King of England (February/March 1626)

The Road Transport Imperative

And to the end all publick officers may be certainly accountable, and no Factions made to maintain corrupt Interests.

Agreement of the Free People of England, 1649

The issues that have been raised in the preceding narrative now require to be examined. In brief, they amount to the following. First, that public inquiries into road and motorway proposals are a denial of law and justice in that persons whose property and livelihoods are affected by them are denied the essential information by which alone they can object to them according to their rights under the Acts. Second, that the road and motorway proposals themselves, with all the immense expenditure involved in them and with all their implications in terms of land-use planning and economic development, are entirely without Parliamentary approval. And finally that this country does not have a Department of Transport, but a *Department of Highways, which possesses the power to make policy decisions on railways and waterways and, furthermore, which exercises this power not in the national interest, but in the interest of one industrial/financial lobby – in short, that it thereby constitutes a corruption of government and thus a major threat to our democracy.*

That something is fundamentally wrong with the body politic of this country is now a commonplace and acknowledgement is everywhere: Parliament no longer has control over expenditure; civil servants hold the real power; our representative system of government has broken down. These three complaints are most frequently heard. They say essentially one and the same thing, and they are all true, as the following quotations clearly demonstrate:

> Parliament has failed to adapt itself to deal positively with the great expansion of central government responsibilities. Accordingly it has been reduced to little more than a talking shop in which M.P.s are mostly restricted to negative criticism of and ex post facto investigation of departmental decisions they have had no chance to shape.
>
> Thirdly, as ministers have become submerged by their responsibilities and Parliament is increasingly by-passed in the formulation of policy, Civil Service power is replacing ministerial and Parliamentary power; civil servants have stepped into the power vacuum left by the politicians. (Lord Crowther-Hunt writing in the *Guardian*)

> I wish to refer briefly to the form, content and scope of the White Paper, which I consider an insult to the collective intelligence of the House. It is wholly inadequate to the crucial and important task of this House of acting as a watchdog on the Executive. The information contained in it is patchy in content, inadequate in detail, inconsistent and ambiguous. It does not provide adequate means for the elected legislature to scrutinise, examine and question the big issues of public policy, the justification for expenditure or its effectiveness . . . The presentation of the White Paper prevents any rational examination of the course of public spending over the next five years . . . (Mr John Garrett M.P. in the debate on the Public Expenditure White Paper, 9 March 1976).

It may, of course, be doubted whether our democratic and representative system has ever 'worked' in the full or ideal sense, but this in no way invalidates the need to identify what is wrong now and to do our best to put it right. There is an important reason why this is so. In the past we could perhaps afford all sorts of inefficiencies and imperfections; *it can justifiably be argued that we can do so no longer, in that the scale of our great technocracies and the effects of decisions made are now so great that, more than any other time in our history, there is need to operate effective checks and balances within our decision making and our system of government.*

* * *

A characteristic of our age is what, for want of a better term, I will call the *technological imperative*. I will seek to define it by listing its component parts. *First* there is the technology itself, which may be road transport or nuclear power, the oil industry or the computer industry, to choose from a long list. The *second* component, inextricably linked with the first, is an industrial/financial complex, transnational in character, the role of which is to invest in research and development in, and to build, sell and use the technological hardware; in the case of the road transport imperative this includes the motor construction and distribution industry, the civil engineering companies, the petro-chemical companies, the road haulage industry, the suppliers of road building materials and plant, the demolition companies and the relevant trade unions. The *third* is a 'lobby', an organisation based on the industry, the specific function of which is to ensure (by for instance recruiting M.P.s to its cause) that decisions are made at various levels of government in its favour. The *fourth* component is an 'interest section' within the relevant government ministry or department, the personnel of which have close connections with the industry and whose role is to co-operate with the lobby to help ensure that the favourable decisions are in fact made.

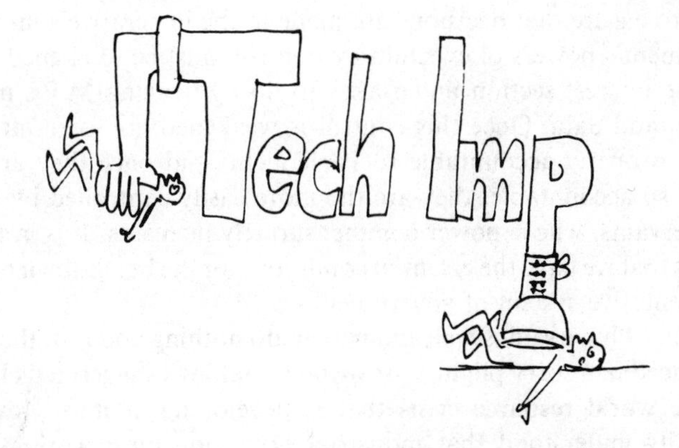

The *fifth*, an obvious component, is a body of expertise in the form of scientists and engineers, all of whose careers are dependent upon the industry. And the *final* component is a brain-washing organisation, loosely staffed by hack economists who can be guaranteed, with endlessly repeated shibboleths obscured in high-sounding technical language, to establish 'economic truths' in the interest of the imperative. Examples of 'truths' of this kind so propagated are:

'Road haulage is efficient, being door-to-door, while rail haulage is inefficient' and 'Electricity generated from nuclear is cheaper than that generated from any other source'. Both are absurd; neither survive a moment's intelligent reflection. It is the work of the economic hacks of a given technological imperative to ensure that intelligent reflection does not take place, or if it does that it is not heard.

It should be clear by now, that with these six components, once a technological imperative is in being, it is practically unstoppable. It decides its own future development by manipulating the government decision-making machinery; it has, by virtue of industrial expansion and proliferating contracts, the power of self-sustaining increase and growth; it is impervious to criticism; and ever-fresh and seemingly undeniable justification for its existence and expansion is provided by the engineers and technologists whose careers and promotion ladders depend upon it.

The importance of one of the components, however, cannot be exaggerated. The interest section within the relevant department is utterly crucial. Linked and cemented to the industry by bonds such as directorships on retirement, it is this factor which authenticates the three complaints or laments with which this chapter began. For in order to ensure that decisions are made in the imperative's interest, Parliament's powers of expenditure control must be weakened, and this the interest section undertakes to do by denying M.P.s information and data. Once this control is weakened, ministers are no longer properly accountable to Parliament; and once they are no longer so accountable, they are the more easily controlled by their civil servants, whose power commensurately increases. It is in these factors that we have the essential conditions for the breakdown of our representative system of government.

If what I have said is true and we can do nothing about it, then we are indeed in a sorry plight. For without making exaggerated claims for the world resource crisis that is developing, it must now be generally understood that industrial expansion on a world scale, endless and uncontrolled, can only lead to major world conflict over increasingly scarce resources.

It is, of course, the object of a technological imperative to expand at its own rate and escape any form of control. Our concern here is with one, the road transport imperative, and with one all-important facet of its development and growth, namely road construction.

More road space and more street space will have to be provided, whatever the plan for it or the cost of it. (Lord Montagu in *The Motor*, 29 March 1927)

As the above quotation makes clear, the nascent road transport industry realised at a very early date the need to control government decision-making in order to ensure the provision of road space. Evidence abounds, showing its early realisation of its other necessary objective, namely the undermining of its rivals in the field, railways and waterways, but this matter is not our present concern. Lord Montagu's brutal words spelled out the need regarding road construction and, implicitly, the need for a lobby and an interest section to ensure the necessary institutional machinery to bring it about. The measure of their success lies in the substance of this book. Because of the structure of the Department and the road lobby influence therein (see the first and second Archway Road submissions, pp. 65, 77), except during temporary periods of economic recession, the road construction programme bears all the characteristics of a technological explosion, having escaped any form of control whatever.

The preceding chapters describing interventions in public inquiries have shown that, in the absence of any other form and given certain conditions, a new ingredient can be introduced to provide at least a temporary check – and that ingredient is people getting in the way, clogging up the machinery, and in peaceful Ghandian fashion jamming on the brakes by all and every form of action short of violence – in a word by what has come to be known as civil disobedience. I am not, however, suggesting that this should be seen as a final answer, but rather something to be used where necessary, and only in the absence of effective checks and restraints which can and must be set into the government machinery.

The M42 Inquiry, the High Court and the Constitution

The principle that each inquiry cannot hammer out a new national policy is not disputable. But it presupposes a national policy sensibly worked out and carrying broad assent. The fundamental problem today is that such a policy can hardly be said to exist.

The Times, First Leader,
30 August 1976

This book began with the M42 Inquiry at Bromsgrove in June 1973. By a strange coincidence that proposed motorway, the inquiry into it and the events that have since surrounded it help to point the way to a solution to the problem that faces us. The Midland Motorways Action Committee rejected the use of civil disobedience; not only throughout their M42 inquiry and others which have followed within their area, but throughout all the years that have covered this narrative they have conspicuously dissociated themselves from every episode. In taking their complaints regarding the conduct of the M42 Inquiry to the High Court (*Bushell* v. *Secretary of State for the Environment*) they put their faith in the law and what might be called the system. *One of the main objectives of a technological imperative, however, and in particular the interest section therein, is to ensure that the law and the system works in its interest*; and so, for example, the Acts relevant to highway construction and inquiries, wherein are defined amongst other things the Minister's powers, are its creation,

no less than the highways themselves. What is more, departmental lawyers always seek to ensure, by the framing of the statutes and by what is colloquially known as 'the Henry VIII clause', that the statutes in practice grant the maximum powers to their minister that Parliament will concede.

A judge in the High Court, the Court of Appeal or beyond, has but one function: the interpretation of the statutes and their application through case law to particular cases, seeking always to ensure in the process that natural justice be not denied. If the imperative has done its work sufficiently well, the overwhelming probability is that any strictly legal interpretation will be in the imperative's interest. The Ex Parte Ostler Appeal Court judgement is a case in point. In the Bushell case, heard by Sir Douglas Frank Q.C., the M.M.A.C.'s case rested upon two propositions. First that natural justice required that objectors had a right to cross-examine all witnesses of the Department on all matters of fact (and that forecasts of traffic growth come within that purview); and secondly that no new evidence should be considered by the Minister subsequent to the inquiry on the grounds that objectors would thereby have no right to challenge it. The case was lost on both counts. From the cases cited in the judgement this can only be regarded as a measure of the successful implementation of the 'Henry VIII clause' concept within a number of ministries and in particular the Department of Transport. Here I require to be less concerned with the judgement itself than with the case for the Department as presented at the hearing by Mr Woolfe, and in particular his interpretation of the Minister's powers under the Acts. (I am grateful here for notes taken by M.M.A.C.)

Effectively, he said three things: *first*, that a Minister *has* power to make a decision, taking account of matters beyond the scope of the inquiry, for he has to consider matters beyond damage to individuals, and to see the problem in the context of a larger whole; *second* that it is Parliament's task to hold a national debate, and that such matters are not fit subject for a local inquiry; and *third* that, though it is one of the fundamental rules of natural justice that a person should not be judge in his own case, a Minister *is in this position in respect of a motorway scheme*.

To take these points in order:

1. Despite the judgement in the case of *Nicholson* v. *Secretary of State for Energy* given by the same judge, some of whose words in the judgement (*The Times*, 6 August 1977) read: 'it used to be commonly thought that the purpose of a public inquiry was to enable local residents and organisations to blow off steam. But that is no longer the case. Persons and bodies opposed to a project now expect to take an active, intelligent and informed part in the decision-making process . . .', no one could sensibly argue

that decisions regarding the building of stretches of motorway could be made by local residents and objectors, *though they may, by the presentation of their objections, compel a re-examination of the proposal by government*. For the decision to build a stretch of motorway or not *must* be part of a national and strategic plan, and must be made beyond the context of persons concerned in defending their property or local environment. Neither in my appeal to the Council on Tribunals, nor in any of my submissions have I suggested that the decision should lie with local objectors, or argued that national policy (the 'context of a larger whole' to use Mr Woolfe's phrase) should be *debated* at local inquiries. What I have consistently argued, however, is that evidence and data concerning that national policy or 'context of a larger whole' should be available to objectors so that they can understand, see the relevance of and need for the proposed road or motorway within that policy, and object to it accordingly.

If we now return to the Government's Consultation Document of 1976, the relevant extracts of which are fully set out in the Winchester M3 Submission (Appendix 3), we read the following:

> Yet by common consent we still lack a coherent national transport policy [The Secretary of State's personal statement]
>
> Transport policy will be properly co-ordinated at the national level only when there is a coherent framework.
>
> By far the most important [dominant theme] is the need to clarify the precise objectives of a national transport policy.

In the face of these statements we require to ask: Where then, is Mr Woolfe's 'context of a larger whole' within which the Minister must see the problem? The answer is that it doesn't exist. It didn't exist when the M42 (or any other motorway) was subjected to inquiry and, as we shall see, it doesn't exist now. And if it didn't and doesn't exist, then the Minister was and is *ultra vires* in taking account of matters beyond the scope of that or any other inquiry; that is, *he could not and cannot possess the powers that Mr Woolfe claims for him*. This is not to say that this could be sustained in the Courts, where, again, the 'Henry VIII clause' may effectively be made to counter it, and we should then be back where we were, appealing to a system organised in its interest by the imperative. Instead, I suggest, objectors should simply refuse the Minister a hearing at his inquiry until he can be seen to have that 'larger whole' from which he can justify his scheme.

2. Mr Woolfe's second point is that it is Parliament's task to hold a national debate and such matters (the 'context of the larger whole') are not fit for a local inquiry. Quite so. It *is* Parliament's historic role to hold such a debate. But the fact is that Parliament has *not* held such a debate; it has been *denied* such a debate (see Hansard extracts, p. 153). With regard to the Secretary of State's 1971 Road Strategy Paper initiating the major road network of which all currently disputed roads are but component parts, *Parliament was denied a White Paper let alone a debate.* The present incumbent of the office, Mr William Rodgers said, most helpfully, in the debate on the Consultation Document in January 1977: 'We have waited for this major debate for nine years.' And the last incumbent said in January 1976: 'No systematic Government statement on Transport Policy has been made since 1968' (Mr Anthony Crosland, *The Times*, 15 Jan 1976). Both statements are true and in themselves dispose of Mr Woolfe's point. But the implications go beyond this. *What this means is first that during those years some 2000 miles of motorway and trunk road have been constructed without any such debate or authorisation of expenditure; and second that under no circumstances can the M42 or any other proposed road or motorway be said to have been the subject of Parliamentary debate or subject to any Parliamentary control any more than the roads already constructed.*

   It should be clearly understood that Parliament cannot effectively debate matters without information, facts, data. Readings of Hansard, exemplified by the extract from Mr John Garrett's speech in the Expenditure Debate, reveal that substantial parts of many debates consist of M.P.s complaining that they *do not have the information on which to hold a debate.* The Select Committee for Expenditure (Transport) Report of 1 May 1975 stated the following:

   We still do not have the equipment, the yardsticks to compare various kinds of transport investment.

   In the debate on that Report Mr Leslie Huckfield M.P. stated:

   I wish that at least similar comparisons could be made between road and rail . . . Unfortunately, since we do not have the same figures, we cannot even make the necessary comparisons which would enable us to decide whether the Department is apportioning expenditure between road and rail on an optimal basis.

Both these quotations bring us to the nub of the issue and to what is and what is not fit subject for a local inquiry. But they do more, for *no plainer words could be uttered on the denial to Parliament of its historic function of control of expenditure. The absence of comparative figures denied and denies it the right to authorise expenditure for the road programme no less than a declaration of Stuart absolutism.*

There being no means of comparison, no coherent framework could have existed by which transport could be properly co-ordinated at the national level. What is not properly co-ordinated cannot, as the Consultation Document acknowledged, constitute a policy. Such was the situation two years ago. The White Paper of June 1977 makes it clear that we are still without such effective comparison:

It has not been possible to develop general rules to make these different techniques directly comparable. The government has asked the independent Advisory Committee on Trunk Road Assessment (the Leitch Committee) to consider the question of comparability of these assessments in its report which is expected later this year. (Chapter 2, para. 67)

So we are no nearer to a transport policy (the 'context of a larger whole') now than when the Secretary of State told us we didn't have one in 1976. *Thus, neither for the M42 nor for any other motorway or trunk road proposed or built during those nine years was there or is there any such thing as policy.*

Now to come back to inquiries into these proposed roads and to the Rules governing them authorised by the Lord Chancellor in 1976. From para. 11.2 we read: ' . . . the appointed person [the inspector] shall disallow any question which in his opinion is directed to the merits of Government policy'. And when objectors, in seeking to establish what questions will and what will not be disallowed (so that, for instance, they will be able to comprehend the need for the road as set against the under-used railway line beside it and object to it accordingly) ask the question: 'What is policy in respect of this road?', they receive the cynical answer: 'The government's policy is to build a strategic network of roads.'

Recourse to the Courts under these circumstances can clearly provide no remedy. For the moment the situation is neither more nor less than a power struggle: the power of the roads imperative represented by the interest section within the Department of Transport *versus* whatever power can be summoned against it. And the only power available is people, ordinary people taking their future into their own hands and simply refusing to allow the inquiry to proceed until that 'coherent policy', that 'context of a larger whole' (*without which neither the Minister nor anyone else can justify the need for the road*) can be seen to exist.

3. Finally, Mr Woolfe claims for the Minister the position of judge in his own case. This large claim can only be sustained given three conditions. First, that the Minister in his proposal is fully answerable to Parliament, and we have seen that this is not the case. The second is that his proposal should be seen against some clearly defined policy, available to all to understand; and we see that this condition is not fulfilled either. And the third is that the Minister's proposal should be the result of impartial judgement by his civil servants in the national interest. A reading of the first and second submissions made at the Archway Road inquiry will establish that this is not the case either. The structure of the Secretary of State's Department and the manner in which decisions are made therein make any impartial judgement in the national interest impossible; the Minister is subject in all his decisions to the most profound bias in the roads imperative interest.

To claim under these circumstances that the Minister can be judge in his own case is demonstrably absurd and a grotesque denial of justice. But from the courts' interpretations so far we are left with no remedy save one – to deny the Minister a hearing, to deny his inquiry validity and to oppose by every peaceful means the opening of what can only be regarded as a cynical mockery of justice and a proceeding wholly corrupted in the road imperative's interest.

The Breaking of the Imperative

Wee are well assured, yet cannot forget, that the cause of our choosing you to be Parliament men was to deliver us from all kind of bondage.

Richard Overton, 1649

Some time in the early spring of 1977 I attended a transport conference organised by the Islington Society. George Cunningham M.P. was one of the speakers, and it was interesting to hear a Member of Parliament describing the problem as he saw it. 'The weakness of Parliament in transport matters,' he said, 'is that there is no body within it responsible specifically for transport and transport expenditure. The Public Accounts Committee looks at past expenditure. The Expenditure Committee is responsible for all expenditure. The Statutory Instruments Special Committee is responsible for all matters. What is required is a select committee on transport.'

It so happened that at this time and quite independently the Transportation Working Party of the Conservation Society had set itself the task of proposing a complete reorganisation of the whole decision-making process in transport. This introduces two major innovations (see the flow chart): a Transport Directorate and a Select Committee on Transport. All the separate (*and equal in status*) transport boards would feed into the Transport Directorate their proposed contributions to the policy proposals in the White Paper. The Directorate would then be responsible for a major procedural innovation, the National Transport Policy and Programme (N.T.P.P.). The Parliamentary Select Committee would keep the work of the Directorate under continuous review, and both bodies would come together in the Examination in Public of the National T.P.P. Transport user groups, pressure groups and statutory bodies would have access to the Select Committee and play a major part in the Public Examination.

From the examination of the N.T.P.P. would emerge the issues which would produce the Transport Finance Bill, which, subject to full Parliamentary debate, would give rise to an Act authorising expenditure on rail, road, waterway, pipeline, freight and public transport in the fields of maintenance or new construction. The individual transport boards would then be responsible for translating the funds authorised into specific proposals, which would become the subject of local inquiries and ultimate action. As proposed, the process would take three years. Parliament would then debate a formal Report from the Select Committee and the process would commence all over again, with a new set of policy proposals, ideally influenced by the Select Committee's Report and the debate upon it. Constant review would be necessary and the power to propose amendments and modifications to the process would be a natural responsibility of the Select Committee.

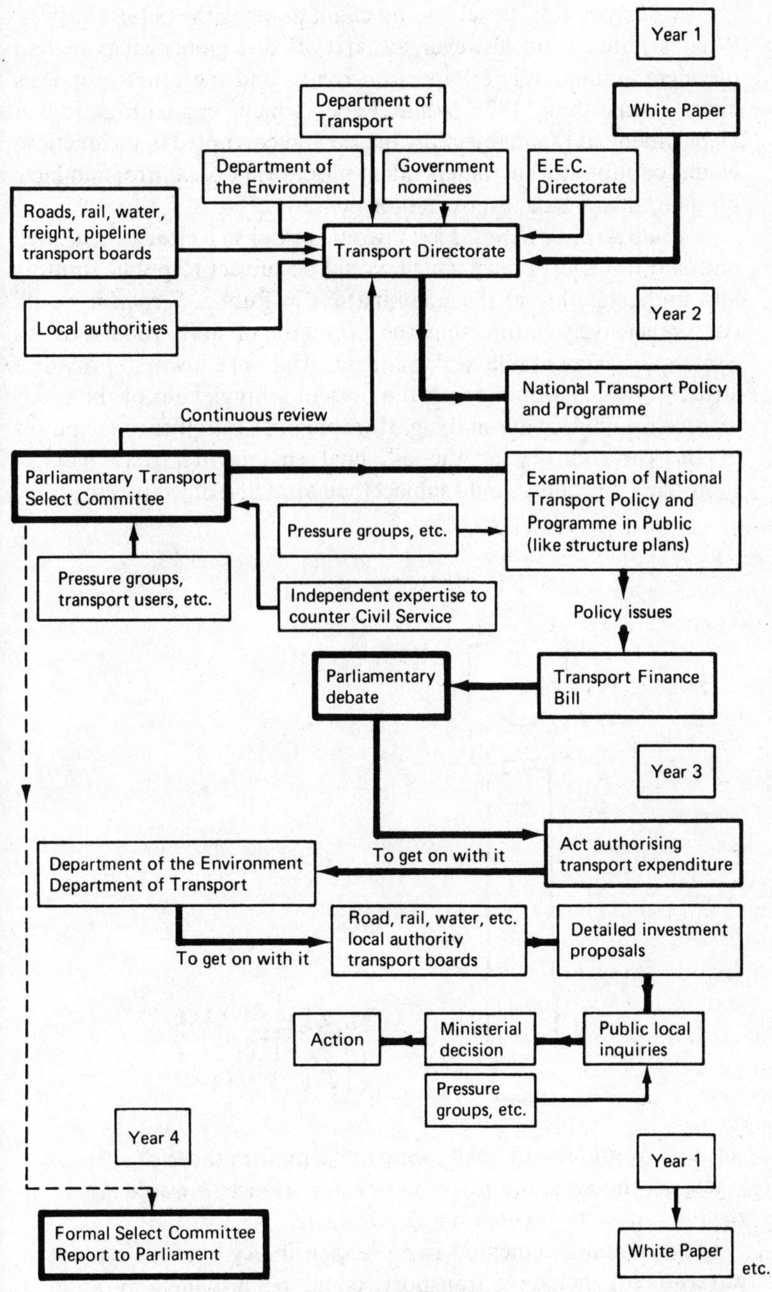

Proposed reorganisation of the decision-making process in transport

Obviously this proposal lays no claim to being the perfect answer. What is quite certain, however, is that it will do a great deal more than the mere 'annual White Papers on roads' and the other proposals made in the June 1977 White Paper which, emanating from a Department still fundamentally biased and corrupted in its function, would be subject to no independent monitoring or control and have no guarantee whatsoever of implementation.

It would do three things: *First* it would establish a clearly delineated national transport policy which would be subject to public scrutiny and understanding at the Examination in Public. *Second* it would comprehensively ensure that the Secretary of State (and thus his officers) be accountable to Parliament. And *third* it would provide a structure which would break the present stranglehold of the roads imperative on decision-making. It would not of itself curb the power of the car industry or the oil, civil engineering, road haulage industries, etc. but it would subject their aims and objectives to public

scrutiny, would reduce their lobby to parity with those of alternative transport modes and, of paramount importance, *it would dismantle the interest section within the Department*.

Further reinforcement, having responsibility over all aspects of government, including transport, could be provided by another constitutional innovation along the lines suggested by David Henderson in his radio talk (Radio 3, November 1977) 'The Unimportance of Being Right'. His suggestion is for a radical and entirely independent body 'in no way bound to observe the established conventions and etiquette of Whitehall' and with only one vested interest – that of ascertaining the truth. It would be a powerful investigatory committee, independent of any minister; its staff would not be subject to the Official Secrets Acts; it would be responsible directly and only to Parliament; and, adequately if not lavishly staffed, it would provide the Expenditure and Select Committees with the necessary expertise and access to information and documents.

* * *

The title of this book implies that democracy is endangered by the overweening power of the roads imperative and its successful subversion of Parliament's powers. There are other dangers. But it is no use complaining that our parliamentary and democratic system of government was never conceived to handle the problems presented by our technocratic age. That is true; it wasn't. It was conceived to provide, amongst other things, an alternative to Stuart Absolutism, to Royal Monopolies and to the Divine Right of Kings. A new kind of 'bondage' from which we require to be delivered has emerged, that of Bureaucratic Absolutism, the Technocratic Monopoly and the Divine Right of the Technological Imperative. Parliament can be adapted to deal with that. All that is necessary is the will.

The corruption of the process of decision-making in transport has led to a dangerous erosion of respect for the rule of law amongst large numbers of people; this is no longer a matter for dispute and should be cause for growing alarm. That I have been associated, as this book makes clear, with the early and peaceful manifestations of protest, I am proud to acknowledge, and I shall continue to involve myself to the full. I shall do so in the hope that the necessary reforms and changes can be brought about in time; that is, before the roads imperative takes us beyond the point of no return, and those ugly forces which are never far below the surface emerge to make capital out of the nationwide frustration and despair that would otherwise be created. I am of the opinion that there is little time left to us.

All that is necessary for the triumph of evil is that good men shall do nothing.

Edmund Burke

Appendix 1

Correspondence between the Conservation Society and the Council on Tribunals

Conservation Society
Transportation Working Party
66 Brookhouse Hill
Fulwood
Sheffield S10 3TB

21 December 1973

The Secretary
Council on Tribunals
6 Spring Gardens
London SW1A 2BG

Dear Sir

I am writing to you on behalf of the Conservation Society to ask for the Council's help in the matter of motorway public inquiry procedures, in particular, the inquiry (opening date 22 January 1974) into the M56 (North Cheshire) proposals, made on behalf of the DoE by the North Western Road Construction Unit, Preston.

1. The Conservation Society is acting as an Objector at this inquiry. In writing to the Director of the NWRCU to request information, I stated that, in the light of the entirely new situation that has developed since the general road strategy was originally presented to Parliament in 1971, it will now be necessary to relate his proposals to the whole new field of integrated transport planning that is now beginning to emerge from the DoE, together with the long-term restraints on road vehicle movement that now seem inevitable.
2. The reply, which I have just received, states: 'It is not appropriate

for the Department to debate these policies at a public inquiry. The department would not, therefore, propose to give evidence in support of these policies or to seek to rebut arguments about the merits of them by objectors . . . You will see that it is not intended to relate the proposals to an integrated transport policy in the manner you suggest.'

3. Our purpose in writing to you is to express our most strong objection to this position adopted by the R.C.U. We put it to you that to open this inquiry without any reference to the new situation (in particular on the matter of recently declared Government policy on urban traffic restraint and increased rail investment with moneys taken from the road programme) and to refuse to relate the proposals to it, is to render the inquiry *ab initio* invalid, in that the proposers have stated their intention to present evidence which can have no relevance to the present economic, planning and energy situation, and which, therefore cannot be anything else but meaningless. We put it to you that this effectively prevents any proper public understanding of the scheme proposed and that, therefore, the inquiry should not be opened on these terms.
4. We, therefore, make to you the strongest representation to secure a change in policy whereby public inquiries into motorway proposals shall be rendered meaningful and comprehensible to the public and the individual objector. And we request, in particular, your intervention in this current matter in order that the N.W.R.C.U. can make the necessary alterations to their present proposals.

Yours faithfully
John F. Tyme (signed)
Counsel for the Conservation Society
at the M56 Inquiry

Conservation Society
Transportation Working Party
66 Brookhouse Hill
Fulwood
Sheffield S10 3TB

2 January 1974

The Secretary
Council on Tribunals
6 Spring Gardens
London SW1A 2BG

Dear Sir

Thank you for your acknowledgement of my letter of 21 December. The purpose of my writing now is perhaps to further elucidate the appeal that we make to the Council, as follows.

My attention has recently been drawn to the Department of the Environment's recent Paper on Public Participation and an extract therein that is highly relevant to the point that we are seeking to establish.

This reads:

> So far as concerns questioning of policy, the determination of transport policy on a national level is a matter for the Government to keep under review and to decide, subject to debate in Parliament. It is not appropriate for the general underlying policy to be brought into question in relation to the various stages of development of individual schemes.

The point here that I wish to establish is that bringing the 'general underlying policy' into question is not the issue. What is at issue is that we, as objectors, wish and need to know in what way this individual scheme proposed by the N.W.R.C.U. has taken account of that 'transport policy on a national level . . . subject to debate in Parliament' which exists now, or at the opening of the enquiry in January 1974.

As it stands at present, this proposed scheme makes good sense and we can understand it in the light of Government transport policy in the early 1960s. It makes sense and is comprehensible to the public and the objector in the light of Government transport policy in July 1971, the date the 'road strategy' was announced to Parliament. But it doesn't make sense and it is not comprehensible to the public and the objector in the light of Government transport policy now, in January 1974.

For since July 1971 a quite major change has taken place in Government policy. This is exemplified by Government statements in response to the House of Commons Select Committee on Expenditure, where it acknowledges the need for a change in policy in urban and connurbation areas whereby traffic must be restrained and public transport increased. It is exemplified by Government announcements on general transport expenditure which mark a radical departure from previous policy in that substantial funds are to be withdrawn from road expenditure and transferred to rail. It is exemplified by the statement of the Minister for Transport Industries that he is willing to consider allocating funds for the building of private rail sidings to link firms directly with the rail system. And, of course, a far more massive change is implied in Government statements in response to the increase in energy costs and the need to conserve petroleum. Thus a major shift has been made away from a system in which road transport was seen as increasingly dominant, and in which rail, canal and public transport were allowed to decline, and towards a system of road traffic restraint, public transport increase and transport integration.

But at the inquiry, planned to open on 22 January, the N.W.R.C.U. intend to ignore all this, and to propose their scheme as though it had never happened – 'It is not intended to relate the proposals to an integrated transport policy.'

We submit that in order that this inquiry can proceed within the meaning of the law, and within the spirit of the Department's declared participation procedure, what we need as objectors is precisely the same right to understand the scheme proposed as objectors had in the case of a scheme proposed in, say, August 1971. At that time they may have disagreed with it and thought the scheme and the whole policy wrong, but they could understand it in the terms of the DoE Paper on Public Participation, in that it clearly represented and fitted in with Government policy approved by Parliament. We, on the other hand, cannot understand this scheme, nor can we see the proposed inquiry having any validity, in that the evidence in support of the scheme does not represent and does not fit in with Government policy approved by Parliament.

It is on these grounds, Sir, that we ask the Council's intervention whereby the Department's proposals can be appropriately amended.

Yours faithfully
John F. Tyme (signed)
Counsel for the Conservation Society
at the M56 Inquiry

Council on Tribunals
6 Spring Gardens
London SW1A 2BG

18 January 1974

Conservation Society
Transportation Working Party
66 Brookhouse Hill
Fulwood
Sheffield S10 3TB

Dear Sir

I refer to your letters of 21st and 31st December about the extent to which Government policy is open to debate at public inquiries such as the inquiry into motorway proposals to which you refer.

It may be of help to you to set out the Council's position in this matter. The Council endorse the view taken by the Franks Committee on Administrative Tribunals and Enquiries in 1957 that, whenever possible, some indication of the general policy relevant to the particular proposal under inquiry should be available before the inquiry. They consider that a witness or witnesses of the responsible Department should be present at the inquiry to explain the policy applicable and to answer questions in elucidation of it, showing if necessary how the proposal fits in with that policy; but that departmental witnesses should not be called upon to defend the merits of Government policy for which the Minister concerned is answerable to Parliament.

In response to our enquiries the Department of Environment have informed the Council that there is no change in the general policy of building a strategic network of modern roads and that recent Government statements to Parliament do not affect that policy. I hope this clears up any misunderstanding of the position that may have been left after your correspondence with the North Western Road Construction Unit, which we have seen.

Yours faithfully
W. S. Carter (signed)
Secretary

Appendix 2
The Aire Valley Submission

Procedural Submission

This concerns the matter of the prior need before such an inquiry can open for the Department itself to conform to the law, to wit, the 1959/1971 Highways Acts, and properly to complete the statutory procedures therein laid down.

I should like to illustrate the importance of this matter, Sir, by quoting from statements by senior officers of the Department in this Expenditure Committee 'Minutes of Evidence'. The subject is control of road programme expenditure, and the officers are giving evidence before the House of Commons Expenditure Committee (General Sub-Committee) on 12 November 1973.

In Para. 10, entitled Firm Programme, we read:

If on the basis of the report the scheme promises to give value for money, it will be included in the firm programme for authorisation to start work in a particular target year, *subject to satisfactory completion of statutory procedures . . .*

And in Para. 68, in answering a question by Mr Loveridge, M.P. relating to the possibility of reducing the length of time taken to 'get a road going', Mr Pelling, for the Department, said in his reply these significant words:

The main problem is *the protection of the rights of the individual through the statutory procedures. Ministers naturally must be concerned to protect those rights, just as much as they are concerned to put the road programme through and reap the benefits.*

Now, Sir, I need hardly emphasise that, in quoting those words, I am merely seeking to state the obvious: that, in proposing a scheme and proposing to hold a public inquiry into it, the Department must conform to the statutory requirements, or, to put it more simply, it must obey the law.

It is the substance of this submission that the Department in this instance has not done so, and to that end, may we refer to the 1959/1971 Highways Acts, the First Schedule, Part II, Schemes under Section 11, Para. 7?

Here we read:

> Where the Minister (a) proposes to make a scheme under Section eleven of this Act . . . the Minister . . . shall publish in at least one local newspaper circulating in the area in which the special road to which the scheme relates is situated a notice (a) stating the general effect of the proposed scheme.

Now, Sir, what is meant by 'general effect'? I should say first that we had no guidance in the interpretation of this phrase prior to the Town and Country Planning (Consolidating) Act 1971. In Part II, Section 7, Subsection 4, on 'The Formulation of Structure Plans', we read:

> In formulating their policy and general proposals under subsection 3(a) of this section [subsection 3 states that the structure plan for any area shall be a written statement and, of course a published statement] the local authority shall . . . have regard: (a) to current policies with respect to the economic planning and development of the region as a whole; (b) to the resources likely to be available for the carrying out of the proposals of the structure plan . . .

And so, with respect, we submit that, just as it was clearly the intention of Parliament when drafting the Town and Country Planning Act that the public should be informed of something of the general effect of the structure plan proposals with respect to the development of the region as a whole, so Parliament when drafting the Highways Acts clearly intended that the Minister should inform the public by means of a published notice in a locally circulating newspaper something of the *general effect* of the proposal with respect to the development of the region as a whole, showing, in the words of the Act (Section 11, Subsection 6), that he has given 'due consideration to the requirements of local and national planning, including the requirements of agriculture'.

And may I now say, Sir, that what the *general effect* of this proposal is, in the mind of the Department, is for them to decide. *Our* view of course of the general effect of this proposal is no doubt vastly different from theirs. That is not germane. Whatever *general effect* this proposal is to have upon this area in their view *they are required to state in the public notice*. However briefly they do so, and however *general* they view the effect within these terms is again up to them. They may restrict themselves to the general effect upon transport modes in the area, to changes in land-use, or to general economic development. That is for them to decide. It is a simple matter whereby they are required to put before the people of this area something of the general planning effects of the proposal.

And so, Sir, to get back to Schemes under Section 11, Para. 7, has the Minister published in a local newspaper a notice telling the people of this area of Yorkshire the general effects of this proposal as they relate to the development of the area? He has not. What he has published, ostensibly in response to that statutory requirement, is a notice a copy of which I have here. And I respectfully submit that from reading this notice no member of the public could glean the remotest conception of any effects of this proposed scheme, be they general within any meaning of that word, or specific within any meaning of that. The notice does not even mention the word 'effect'. It is merely a notice of an intention to make a scheme, and to make an order. That, and no more.

At this point may I anticipate a possible line of argument, and in doing so refer again to Schemes under Section 11, Para. 7?

It has sometimes been argued that the Minister *has* fulfilled the mandatory requirements of the para. in that, in publishing the notice that we have here, in naming a place where a copy of the draft scheme, plans, etc. may be inspected, and in stating the right to object, he has published the general effect of the proposed scheme.

Now, Sir, though I find it difficult to understand how this view could seriously be held for a moment, I think I should dispose of it here and now by a close examination of what it would seem to mean.

What the assertion means, if it means anything, is that by complying with Para. 7 (b) and Para. 7 (c), the Minister has complied with Para. 7(a).

It should be unnecessary to point out, Sir, that if that *were* the case, then the drafters of the Act would have written the para differently. They would have written:

> Where the Minister proposes to make a scheme . . . he shall publish . . . a notice stating the general effect of the proposed scheme:

(a) naming a place . . . etc.;
(b) stating that . . . etc.

But they did not write that. They wrote:

Where the Minister proposed to make a scheme . . . he shall publish . . . a notice:
(a) Stating the general effect of the proposed scheme . . .

And this, of course is what they have not done

* * *

Now, Sir, in that the Minister has not conformed to the statutory requirements, what follows? The requirement, you will note, is mandatory, 'shall' publish. A precedent on this matter was given in 1970 in a matter relating to the Town and Country Planning Act, Section 88 (2) 'Appeal Against Enforcement Notice', which reads:

An appeal under this section *shall* be made by notice in writing to the Secretary of State, which *shall* indicate the grounds of appeal and state the facts on which it is based.

When, Sir, solicitors failed in this mandatory requirement in that they indicated the grounds but did not state the facts, the *Journal of Planning Law* (1970) p. 431, reported that the Secretary of State refused to entertain such appeals. He said, 'I have no jurisdiction'.

By any interpretation, Sir, this means that you have no jurisdiction to open or preside over this inquiry. A mandatory statutory requirement has not been complied with by the Department. This renders the inquiry a nullity until it has.

And may I now, Sir, with your permission, anticipate a reply from learned counsel across the floor, and save us all time? We do not accept his invitation to apply to the High Court for a writ of Mandamus. This is not, Sir, some subtle matter of interpretation. It is a clear case of dereliction. We have here a clear *prima facie* case of failure to comply with a mandatory requirement upon which you must now decide. Then, Sir, when you have upheld the law and suspended the inquiry to allow the Department to fulfil its statutory requirement, it is then, of course, open to the Department to apply to the High Court themselves.

You see, Sir, to conclude, we require to see this inquiry fulfil Mr Pelling's well chosen words. Ministers, he said, must be concerned to protect the rights of the individual through the statutory procedures. You are responsible, Sir (and particularly in this day and age when, to quote the Master of the Rolls on October the eleventh of this year: 'The Rule of Law is in greater danger today than it has been for centuries'), you are responsible for ensuring that the rule of law will be upheld in this inquiry, no matter with what contempt it has been treated in other inquiries. In the case of Schemes under Section 11, Para. 7, it has not been upheld. Until the Department has complied with that mandatory requirement, you have no jurisdiction to open or preside over the inquiry.

Accordingly, I am requested in this submission to ask you, in that it is your duty to protect the rights of the objectors here this morning through the proper observance of the statutory procedures, to declare this inquiry here and now null and void.

Paragraph 7, Sub-paragraph (A), First Schedule, 1959 Highways Act

Note on interpretation We have read the Department's interpretation of Para. 7 (transcript of proceedings at M16 inquiry, Epping, Mr John Newey Q.C. for the Department). Here it is claimed that the word 'scheme' is defined in the statute as an instrument empowering the Minister to make a special road; from which it follows that the general effect of the instrument is – the special road; from which it further follows that, in publishing the disputed notice in which he delineated the route of the road, the Minister has complied.

We reject this on the following two grounds. First, it requires a narrow and obscurantist interpretation not merely of the word 'scheme', but of the whole para; an interpretation which would be utterly beyond the reasonable comprehension of the common man. Worse, it denies the para any meaning. We deny that the GENERAL EFFECT of the instrument can be the FORM of the instrument. For the road is the *form* that the instrument takes as a result of the Minister's action; the road is not the general effect of the instrument and never could be. The para requires 'the general effect of the scheme', that is, the general effect of the form that the scheme (a mere piece of paper, to quote Mr Newey Q.C.) takes as a result of the Minister's action; thus the para requires the Minister to state the general effect of the special road. He has not done so.

Secondly, elsewhere in the schedule, the word 'scheme' as used cannot on any reasonable grounds whatever be interpreted as meaning a mere statutory instrument. Para. 10 empowers the public to object to the 'scheme'. When the Conservation Society has presented objections to a scheme it has not objected to a statutory instrument (that would have given it no grounds upon which to object), but to the form that the scheme takes, that is the special road. In the same way, all objectors act in accordance with their statutory rights under that para. In doing so, they are denying (and the Department clearly is accepting their denial) that 'scheme' as used can mean anything but the road. And if 'scheme' be accepted as 'road' in Para. 10, then it must similarly be accepted in Para. 7.

The Minister has not complied.

The Matter of Rules

I wish to deal here with the matter of the absence of Rules of Procedure. I think I should make it clear that I have raised this matter at a number of motorway Inquiries, including the M16 at Epping last December. There, in my submission to Mr Clinch I drew attention to the following:

1. That there are no Rules established for inquiries under the Highways Acts;
2. That there are Acts empowering the making of such Rules;
3. That the holding of such inquiries without established Rules, available to all parties, is contrary to the Law of Natural Justice in that it leaves the public without protection, in support of which contention I gave examples; and
4. That it would be wholly improper for that inquiry to open and proceed until the anomalies were removed by the introduction of such properly constituted Rules.

Three things followed from this. First, the Inspector in his opening statement on the second day made reference to this matter, substantially as follows: There is no statutory duty laid upon the Secretary of State to make Rules, but that does not mean that the public are without protection, as he, in carrying out his office, had to obey the Law of Natural Justice. This could then and can now hardly be said to help us, as I shall endeavour to show in a moment. The second consequence was that the Inspector sent this, together with the two other submissions, to the Secretary of State for his comment and answer. And the third consequence was (as recorded in the transcript) that the Inspector stated that we would have an opportunity later in the inquiry to pursue the matters further.

Unfortunately this promise was not upheld, and it is for this reason that we now appear further to pursue these matters before you here in this inquiry. The M16 at Epping is, in our view, the appropriate place but, denied a reappearance there, we have no other alternative but to re-raise these matters here.

* * *

I must now refer to the Secretary of State's address to the M16 Inspector and Counsel for the Department's submission put before him and the inquiry on 9 January to see what they have to say upon the matter, and if I could now ask you to refer to these documents, we find it referred to (and marked accordingly) in Paras 15, 16 and 19 of the former and Paras 10, 16 and 19 of the latter. It is Paras 16 and 19 in both cases which specifically refer to my submission, and from them we learn three things:

1. There are no Rules;
2. There never were any Rules; and
3. The 1958 and 1971 Acts merely empowered the making of Rules; neither required that Rules should be made;

and Para. 19 in both documents tells us, and I quote:

> In the circumstances it does not appear to the Secretary of State that anyone is suffering prejudice at the present Inquiries because there are no procedural Rules under the Act of 1971 applicable to these Inquiries.

But, Sir, how can this be said to help us? As I shall proceed to show you in a moment, it in no way throws any light whatsoever upon the matter that I raised in my submission. We are agreed on point one – there are no Rules. We are agreed on point two – there never were any Rules. We are agreed on point three – the Acts did not require, they empowered. But we utterly deny that Para. 19 has any substance whatever in that neither it nor the preceding Paras made reference to or acknowledged the substance of our submission.

* * *

Our case rests upon the following:

1. That the Department should have used the powers granted to it by Parliament and made Rules; the empowerment was granted for this, not to be set aside and ignored; and
2. That the absence of Rules leaves objectors and the public without protection and DEPRIVES THEM OF THEIR RIGHTS IN LAW.

May I very briefly now refer to the substance of my original submission of 3 December? And in doing so, may I make this point. We can well understand how the Inspector, under the exceptional circumstances of the opening day of an inquiry, was unable adequately to deal with the matter. However, that the Secretary of State has failed to answer it in his considered reply delivered over a month later is surely highly significant. We must assume that he has nothing to say upon it, and we must all, including you, Sir, draw our conclusions from that.

But to return to the submission:

The handbook *Public Inquiries into Road Proposals* (the Department publication which, in the absence of anything else, is the only source of procedural information available to the public) states in its Para. 20 the following:

> But the formal Rules, where they apply, lay down that the Inspector may disallow questions to Departmental representatives which, in his opinion, are directed to the merits of Government policies.

And so, what this must mean is that, when an objector at this inquiry, exercising his rights under the Highways Act 1959, Part II of the First Schedule, Para. 10, which states:

> [The Minister] after considering any objections to the proposed scheme that are not withdrawn . . .

when he objects that the scheme should not be built at all because it is contrary to the national interest (which could hardly be other than 'directed to the merits of Government policies'), and when, in presenting this objection directed to the merits of Government policies, he requires to ask relevant questions of Departmental representatives, you, Sir, may disallow them. (And without information, which only Departmental representatives can give, how can an objector properly exercise his rights under Para. 10 above? And the answer is that he cannot.)

And so, Sir, as things stand, we have the situation that, in following Para. 20 of the *Handbook* you will be denying an objector his statutory rights under the 1959 Highways Act.

* * *

And may I at this point deal with the M16 inquiry inspector's statement on this, which I took down. He said that he would hear anything and everything that objectors might wish to put before him, and *he would have the Department answer all their questions.*

With respect, Sir, we submit that an inspector cannot make such a statement. He does not have the power. You, like us, must be guided (in the absence of properly constituted Rules) by the only rules available to us, and in this matter it is clearly laid down by Para. 20 of the *Handbook*. To say what the M16 inquiry inspector said on 4 December last is to act in contravention of that para. and he is not empowered to do so. In our view this is quite clear; but what must now be clear beyond dispute is that if there was any doubt about this in December last, there can certainly be no doubt about it now. For of course, we look at once in the two documents pronouncing upon this matter coming from the Secretary of State for anything giving you such power, for any reference to Para. 20. And there is none.

Thus, Para. 20 must stand, and you will disallow questions which in your opinion are directed to the merits of Government policies.

But may we now dispose of any lingering doubts on this matter by imagining for the moment that you, Sir, intended to have the Department answer all objectors' questions – in other words to compel Departmental representatives to answer complex questions on, for instance, electrification plans for British Rail in this area, plans for additional and private sidings, plans for increased rail distribution centres, investment plans for Yorkshire waterways, for boat loading facilities, plans for energy conservation contained in the 1974 Railway Act, etc.; then, Sir, in our submission, one of three things would happen:

Either: Counsel for the Department would get up and say that he had no intention whatever of putting up witnesses to answer such questions (as was the case when, before Inspector Farrar in March 1974 I asked such questions) or: The Secretary of State would tell you, through his Counsel, to stop

or: The Secretary of State would, on receipt of your report, refuse to consider any evidence put before you by any objector as a result of your permitting such questions contrary to Para. 20 of the *Handbook*.

In other words, one way or another, objectors will be deprived of their rights under the Highways Act of 1959. Effectively, we have the extraordinary position whereby a matter of your opinion can deny a member of the public his statutory right under the Act. YOUR OPINION IS SET ABOVE THE WILL OF PARLIAMENT – a major constitutional anomaly which must be terminated at once.

We have no doubt that it is because of this, among other matters, that the Council on Tribunals wrote to the Department (as the Secretary to the Council told me in person in May last year) requiring that Rules for Special Roads Inquiries be made. The Department, doubtless for its own purposes, ignored the Council's representation. They have now ignored ours.

The responsibility, in our submission, now, therefore, very clearly lies with you. Your independent status in this Inquiry, Sir, whereby you seek to uphold the law, to see natural justice done, means now one thing. You can have but one course, as we said at the opening day of the M16 inquiry: to adjourn the inquiry in order that properly constituted Rules, sanctioned by the Lord Chancellor's Office, be made out to remove this and other anomalies. For the inquiry to continue would mean, quite apart from adding to the growing public disquiet and cynicism on these matters, that you would be DENYING PARLIAMENT ITS SOVEREIGN POWERS IN LAW.

The Matter of the Council on Tribunals' Ruling

May I now turn, Sir, to the second issue that concerns us, the subject of our second submission at the M16 inquiry, namely the matter of the Council on Tribunals' Ruling?

Again, it is our central contention that the Secretary of State, in the two documents put in by him to the inquiry in no way answered the substance of our submission. We readily accept, again, that the M16 inquiry is the proper place for the consideration of this matter, but as I have already stated, our exclusion from that inquiry makes that impossible, and thus, necessary for it to be raised before you, Sir. I am sure you will understand as a consequence the need at this point for a short résumé of the essence of the original submission.

In briefest form, it made the following points:

1. That there have been significant changes in Government transport plans and policies since the key (and so often referred to) Road Strategy Paper was put before Parliament on 23 June 1971;
2. That this can be acknowledged by reference to a major strategic planning document for North-east Lancs. in 1972 (then quoted);
3. That these changes can be exemplified by *inter alia* numerous statements to Parliament made by Ministers for Transport or Transport Industries;
4. That objectors at motorway inquiries need to know how the relevant motorway proposals fit in with those statements. They need to know, for instance, how the construction of this motorway/extension to the road network helps towards the declared policy of transferring freight from road to rail; how it fits in with the TPPs of the local County authorities; what cognisance it has taken of the decision to link firms with the railway system (for which very substantial grants are now available for private sidings and related plant and buildings); how it complies with Government energy policy, in particular as the 1974 Railway Act was drafted with energy conservation in mind; in what way it might be unnecessarily duplicating the investment of public funds in the waterways of the area; in what way it has taken account of Cmnd Paper 5366, in particular Para. 13 and Government's response thereto. They need to know, in other words, how this scheme can be seen to fit in with what we might term 'general policy';
5. That objectors do not wish or need to debate this policy, but merely to be clear (from Departmental witnesses) as to what it is, and how the proposed scheme fits into it;
6. That until objectors are helped so to understand these matters, the proposed scheme makes no sense to them whatever;
7. That at previous inquiries the Department had refused to help objectors to such an understanding;
8. That the Council on Tribunals had ruled upon the matter in a letter to the Conservation Society dated 18 January 1974 [a copy of the letter was put in – see Appendix 1];
9. That it was only possible properly to construe the Council's letter

and the terminology therein by reference to letters from the Conservation Society to which the Council's letter was in reply;

10. That, in that the Department had failed to carry out prior to the opening of the inquiry the obligations placed upon it by the Council's ruling, and that, as it could only carry out its obligations *during* the inquiry by a transformation of motorway inquiry procedure, it would be wholly improper for the inquiry to open and proceed, and
11. That the matter and the submission should be sent direct to the Secretary of State for his consideration.

What then occurred, Sir, was that the matter was inadequately discussed; we had no opportunity of elaborating our case or answering Counsel for the Department's preliminary response, in particular with reference to Para. 9 above. The inspector did however, dispatch the submission without such explanation and elaboration to the Secretary of State. It is our present case that in his response, he has not in any sense answered the matters raised in the submission.

In support of this, may I now begin by referring to Para. 9 above, and examine the Council's letter, a copy of which you have before you. I wish in particular to draw attention to:

Second para. line 4	'some indication of the general policy relevant to the particular proposal'.
line 7	'to explain the policy applicable'.
and line 9	'how the proposal fits in with that policy'.

Now, Sir, we are concerned here with a matter which the Secretary of State in his response entirely ignored, for the reason that he had no information upon it, but with which we must now be concerned, because it is central, crucial to the matter in hand.

What we are concerned with is this question: What was in the Council's mind when it used the words: GENERAL POLICY; THE POLICY APPLICABLE and THAT POLICY?*

This question can, we submit, only be answered by reference to our

* Here we think we should make plain the clear distinction in the letter between the above references to POLICY and that of the final para: 'the *general policy of building a strategic network of modern roads* . . . recent Govt. statements to Parliament do not affect *that policy*'. That 'that policy' has not been changed is not in dispute. It is 'that policy' that we need to understand in its context.

letters to the Council [see Appendix 1]. You have copies before you, Sir, and may I now draw your attention to certain passages within them?

First, the December 1973 letter:

Para. 1 'it will now be necessary to relate his proposals to the whole new field of integrated transport planning that is now beginning to emerge from the DoE, together with the long-term restraints on road vehicle movement that now seem inevitable'.

Para. 3 'We put it to you that to open this inquiry without any reference to the new situation (in particular on the matter of recently declared Government policy on urban traffic restraint and increased rail investment with moneys taken from the road programme) and to refuse to relate the proposals to it, is to render the inquiry *ab initio* invalid, in that the proposers have stated their intention to present evidence that can have no relevance to the present economic, planning and energy situation, and which, therefore, cannot be anything else but meaningless. We put it to you that this effectively prevents any proper public understanding of the scheme proposed and that, therefore, the inquiry should not be allowed to open on these terms.'

Para. 4 'in order that the N.W.R.C.U. can make the necessary alterations to their present proposals'.

And the January 1974 letter:

Para. 4 'transport policy on a national level . . . '

Para. 5 'But it [the proposed scheme] doesn't make any sense, and it is not comprehensible to the public and the objectors in the light of Government transport policy now, in January 1974.'

Para. 6 'For since July 1971 a quite major change has taken place in Government policy. This is exemplified by Government statements in response to the House of Commons Select Committee on Expenditure, where it acknowledges the need for a change in policy in urban and conurbation areas, whereby traffic must be restrained and public transport increased. It is exemplified by Government announcements on general transport expenditure which mark a radical departure from previous policy in that substantial funds are to be withdrawn from road expenditure and transferred to rail. It is exemplified by the statement of the Minister for Transport

Industries that he is willing to consider allocating funds for the building of private rail sidings to link firms directly with the rail system. And, of course, a far more massive change is implied in Government statements in response to the increase in energy costs and the need to conserve petroleum. Thus, a major shift has been made away from a system in which road transport was seen as increasingly dominant and in which rail, canal and public transport were allowed to decline, and towards a system of road traffic restraint, public transport increase and transport integration.'

Para. 7 'But at the inquiry, planned to open on 22 January, the N.W.R.C.U. intend to ignore all this, and to propose their scheme as though it had never happened – "It is not intended to relate the proposals to an integrated transport policy".'

Final 'It is on these grounds, Sir, that we ask the Council's intervention.'

Sir, we submit to you very plainly that, reading the above extracts from our letters to which the Council's letter was in considered reply, the Council's words 'General policy relevant to the particular proposal', 'to explain the policy applicable', and 'how the proposal fits in with that policy' can only mean with reference to policy – the totality of the Government's transport (rail, canal and public) and energy policy. No other interpretation is possible.

So what we are saying is that the Council on Tribunals has ruled that before the opening of an inquiry into a proposed scheme, the Department should explain the general transport (i.e. road, rail, canal, public) policy relevant to the proposal; and once the inquiry has opened, it should put up witnesses from the responsible department (DoE for transport; Department of Energy for related matters) to explain policy in these matters, answer questions in elucidation of it, and show how this proposed scheme fits in with the total policy. They should not be asked to defend the merits of that policy, merely to show what it is, and how the proposed scheme fits in with it.

Now, Sir, has the Secretary of State complied in this matter with respect to this inquiry any more than he has in the matter of the M16? We submit that he has not. But as, in the case of the M16 a specific claim to the contrary was made and as a reply on the matter was received there from the Secretary of State repeating this claim, may we now examine it?

The matter is referred to in the following:

1. The Secretary of State's statement addressed to the M16 inspector;
2. Council for the Department's submission to him of 9 January; and
3. The House of Commons Adjournment Debate on 21 January (Hansard, p. 1439).

The key paragraphs in the two documents (which you have, Sir) are, in the case of the former Paras 7, 8, 9, 10 and in the latter, Paras 4, 5, 14, 15, 17. If I might briefly and accurately paraphrase them, what they say is this:

1. That objectors can find all they wish to know from documents issued to them before the inquiry opened, and from Counsel's opening statement citing Ministerial statements showing that the motorway has been the policy of successive governments for years;
2. That this particular motorway (M16) is particularly important to relieve traffic; and
3. That the Secretary of State is satisfied that his Department has complied with the expressed views of the Council on Tribunals and the Franks Committee.

In the adjournment debate, in answer to a question from the Hon. Member for Brentwood and Ongar, the Parliamentary Under Secretary helpfully quotes the relevant passages from the prior statement issued to objectors, which I will quote in full:

Government policy was . . .

1. In accordance with statements made in Parliament in June 1971 and endorsed by subsequent administrations to create a national network of trunk roads adequate for the economic needs of the nation, linking major centres of population and industry, ports and airports.
2. To improve environmental conditions by removing congestion and, where practicable, by diverting long-distance traffic and particularly heavy goods vehicles, from towns and villages onto roads suitable for them; and to build roads to standards that will minimise the risk of accident and injury.
3. As part of these requirements, to create an orbital road round London, relieving traffic congestion in the Greater London area by allowing through traffic to skirt the periphery of London. These two schemes from the A10 to the A12, in conjunction with other sections from the A1 to the A10 and from A12 to the Dartford Tunnel will provide an important segment of such a route.

The PUS then went on to say that Hon. Members may be aware that when the inquiry first opened, one of the submissions by the objectors contended that objectors did not know how the proposed M16 fitted in with the Government's general underlying policies. He went on:

> My Right Hon. Friend has responded to this by addressing a statement to the inspector holding the inquiry. In this my Right Hon. Friend commented that the objectors should have referred to the policy statement that I have already mentioned and should have waited until counsel for the Department opened his case at the inquiry, in the course of which he cited reports and ministerial statements to show that the construction of an orbital road for London has been the policy of successive Governments for many years and that its construction as part of a national network of roads, itself part of a wider transport strategy, was the policy of the present Government.
>
> My Right Hon. Friend went on to say that he had been further informed that counsel and witnesses called by him explained the importance of the M16 as part of the orbital road and also by itself, because of the traffic which would use it, and the relief which it would afford to other roads. If objectors had studied his statement and listened to the presentation of the Department's case, they could not have been in any doubt as to what Government policy was, nor how the M16 proposal fitted in to that policy . . .

And so, Sir, what is the position now, five months after the commencement of the M16 inquiry? We are in precisely the position which led the Conservation Society in the first place and almost eighteen months ago to refer the matter to the Council on Tribunals. It was precisely such statements as I have just read out, having nothing whatever to do with anything but the building of roads and motorway networks relieving traffic congestion, etc., etc., that we referred to the Council for its ruling. The position we are now in is *exactly* the situation that the Council ruled upon almost eighteen months ago.

With greatest respect, Sir, unless words have suddenly entirely lost their meaning, what we have here is plain proof that the Secretary of State has done *nothing whatever* to comply with the expressed views of the Council on Tribunals and the Franks Committee. All that he has done is to ignore them.

I am sure, Sir, that you are now able to appreciate the difficulty and gravity of the present position, and very onerous responsibility that lies upon you in this matter. We accept fully, as has been emphasised in documents relating to this inquiry, your total independence of the Department. Inspectors have spoken in the past of their wish to assist objectors, faced as we are with the immense power, expertise and resources of this Department of State.

Well, Sir, we are now very much in need of your assistance. The Department, despite its assertions, has not complied with the ruling of *the only authority to which recourse can be had in matters of dispute regarding procedure in inquiries of this nature.*

If I might quote briefly the Tribunals and Inquiries Act, 1971, which restates the Council's role and duties:

> There shall continue to be a council entitled the Council on Tribunals . . . to consider and report on such particular matters as may be referred to the Council under this Act . . . with respect to administrative procedures involving, or which may involve, the holding by or on behalf of a Minister of a statutory inquiry, or any such procedure . . .

And so, Sir, in that the Department has not complied with the Council's ruling in that, prior to this inquiry's opening it has given no information on general policy (but has restricted itself to specific information on highways policy), we ask you to declare an adjournment to enable it to do so; and thus to enable you, Sir, if you feel that it is appropriate, to refer this matter yourself to the Council for guidance. That is what it is there for; that is what Parliament set it up to do.

Not to adjourn it, Sir, can only have the inevitable result of increasing the now nationwide public concern, disillusion and cynicism regarding the manner in which these inquiries are held. Thus we appeal to you to see natural justice done, to uphold the law and protect the rights of the private citizen set against a leviathan Department of State by granting an adjournment for the due and proper settlement of this matter.

On the Secretary of State's Failure to Take into Account National Planning

I wish to refer, of course, in this submission to the Highways Act 1959 Section 7, Subsection 2.

I must be brief here and tell you that it is my clients' submission in this instance that the Secretary of State has failed to comply in this matter.

They say, first, that any consideration of the requirements of national planning could only have been given by submitting the Road Strategy Paper of June 1971 and all major subsequent decisions affecting the great highway programme (with its immense national financial, transportation, economic and land-use planning implications) to the rigours of full Parliamentary debate, vote and expenditure approval and control (and not leaving it to mere supply day debates or the ritualised procedures of expenditure subcommittees whose briefs are inevitably restricted).

But they say more: they say that the Secretary of State has not even taken into account, given any consideration to the requirements of national transportation planning, let alone any other mode of national planning. And to that end, they ask me to draw your attention to several matters, drawn from the Supply Day debate on 1 May 1975 into the Expenditure Committee (Transport) Report, from which I must now quote:

1. Mr James Boyden referred to Para. 13 of the Report, which reads: 'We note that expenditure on canals is nevertheless an ineligible category of expenditure' for Transport Supplementary Grant. 'The Committee', he said, 'recommended that this exclusion should be reconsidered. In Para. 3 of the Fourth Special Report the Department', he continued, 'gives that the brush-off by saying: "Canal expenditures are primarily the responsibility of the British Waterways Board." The Committee', he continued, 'is looking for more vigorous intervention from the Department to achieve integration, and the use of financial controls so that the best available policies . . . are applied throughout the country.'
2. Mr Marcus Fox said: 'I view with suspicion the recently published report of the Government's T.R.R.L. which suggested that, despite everything, car ownership and the use of cars is likely to rise to the end of this century at rates not far different from those anticipated before the energy crisis arrived. I cannot swallow that statement.' And later: 'We have been told that the methods of evaluating maintenance on the economic side are not very well advanced. This is not very helpful for the future when we talk about getting the best value for money.' And: 'There is an interesting section in the Report on investment criteria and transport coordination . . . the words "misallocation" of resources is too weak. The word "waste" might be more appropriate. Whatever we call it, in a number of instances duplication and

piecemeal allocation of funds have led to a situation that should not be tolerated.'

3. Mr Nigel Spearing said: 'Everywhere I go, I am told that no one in the DoE knows anything about canals and waterways.'
4. I have left till the last, Sir, Mr Leslie Huckfield's contribution to the debate as, owing to his special position in the matter of transport planning, it is so significant. Here are some of the things he said:

> The Report hammers home the point I have been trying to make on many occasions – that it appears that we still do not have an overall transport policy. When I see the detail into which the Report goes, such as the eleventh recommendation, making the point that we still do not have the equipment, the yardsticks to compare various kinds of transport investment, I feel that the Report has hit upon one of the central matters lacking in present transport policies.
>
> Many of us are still waiting for the emanation from Marsham Street of the White Paper which might set out the Government's intentions on transport.
>
> When we come to the inter-urban or inter-city investment which is made in transport infrastructure, we may come to the conclusion that one part of the Department does not know what the other part is doing.
>
> Just as there appears to be no national co-ordination of our transport infrastructure or investment policies, there appear to be no local policies.
>
> We should be planning for the time when the economy picks up. . . . We cannot do that as long as we do not even compare different transport investments.
>
> I do not believe that [the Department] has done any co-ordination of even the pricing and investment policies of the industries for which it has direct responsibilities.
>
> The Department does not even seem to be co-ordinating those parts of the transport industry and transport expenditure for which it has complete responsibility.

Sir, I submit that everything here cited, quite apart from revealing the total ignorance in which the House is kept regarding transport policy, reveals the *total absence* of anything remotely recognisable as national transport planning. The Act requires the Minister to take account of national planning. He has not done so. He has not complied. The inquiry cannot therefore open until he has.

On the Secretary of State's Failure to Take Account of the Requirements of Navigation over the Waters

Now, Sir, turning to Para. 1 of the disputed Public Notice, we read: 'An Order under Sections 7 and 20 of the Highways Act 1959 . . . '.

It is Section 20 I wish to refer to, which begins: 'The Minister shall take into consideration the reasonable requirements of navigation over the waters . . . '

And my clients require me to ask this question before you: Has the Secretary of State taken the reasonable requirements of navigation over the waters of this Aire Valley canal into consideration? And they require me at once to answer it and say: No, he has not, for the simple reason that, as has already been indicated in part, he has no means of doing so. In further support of this, however, I have two matters to raise before you.

1. One of those thirty-six organisations across the country whom I represent before you today is the Transport on Water Committee (T.O.W.) of the Royal Albert Dock in London. Now, Sir, they put up an Objection at the M16 inquiry at Epping in March of this year, and they took a transcript of the questions they put to the Department and the answers they received. Here are one or two of them, Sir:

   T.O.W. May I ask the Department what plans there are for enlarging the locks on the river Lea and general upgrading from the River Thames to Bishops Stortford?

   Dept. We cannot answer.

   T.O.W. Had the Ministry investigated the GLC's 15 October 1974 survey of the upgrading of the canal from the River Thames to Watford to take 400 ton containers, which stated the cost as being £1m per mile?

   Dept. We have no reply.

   T.O.W. Would you please inform me of the cost of the Lea Navigation between Poplar E14 to Waltham Abbey (14 miles)?

   Dept. If someone says something three or four times, it might make an impression. It is perfectly true we can obtain information from British Waterways, no doubt.

   T.O.W. It is essential to the case that this information be provided from some such source.

   Dept. We suggest that T.O.W. should put forward this information.

Now, Sir, you will readily recognise that this sort of thing is *precisely* what was in the mind of the Council on Tribunals when it stated that the Department should put up proper witnesses to answer just such matters. But that, of course, is in parenthesis. The point I am making before you now on behalf of those people in this valley who live beside this totally neglected canal is that the Department have no knowledge at all (as Nigel Spearing M.P. said in the House) on canal and water transport, either generally or specifically.

But to my second point:

2. A written question was submitted and answered in the House on 7th May this year:

> *Mr Spearing*: Asked the Secretary of State if he will make it his policy to fund capital improvements or additions to commercial waterways traversed by public or privately owned cargo or passenger carrying craft in the same manner as capital for roads; and if he will give the reasons why these forms of common track are not financed in the same manner as at present.
>
> *Mr Mulley, Minister for Transport*: His answer ended with the words: 'The Government contribute, in grants, some 70 per cent of the costs of the British Waterways Board.'

Mr Price M.P. subsequently queried this statement and in a letter, Mr Howell, Minister for Sport (who also is responsible to the Department of the Environment for waterways), wrote to Mr Spearing as follows:

> Frank Price has written to me to query this figure, pointing out that, in fact, grant-in-aid accounts for a considerably smaller proportion of the Board's total costs. This is correct. I fear that we were guilty of over-compression in answering your question. 70 per cent is the proportion which the grant-in-aid approved for 1975 bears to the Board's estimated expenditure on the waterways in 1975 and not to their total expenditure.

So, Sir, here we have the Minister of Transport on 7 May, advised as he is by his Departmental civil servants, misinforming the House regarding so simple a matter as the financing of canals and waterways. How well, how precisely, does this bear out Mr Spearings words in the 1 May debate.

And so, Sir, my clients ask me to submit to you, that under these circumstances, to claim that 'the reasonable requirements of navigation over the waters' of this valley canal have been taken into consideration by the Minister is a palpable absurdity. The total ignorance of the Department on all waterway matters, be they local or general, is self evident. Para. 20 requires the Minister to take into consideration the reasonable requirements of navigation over the waters. He has not complied. You, Sir, have no powers to open the inquiry until he has.

On the Failure of the Secretary of State to Ensure Adequate Parliamentary Control of the Road Programme

Sir, if we look at the Secretary of State's Statement of Policy, sent to all my clients, we read that these matters are, and I quote, 'for decision by the Government, subject to the control of Parliament'.

This, Sir, is a theoretical view with which my clients have no difficulty in agreeing. But what concerns them, Sir, is not the abstract constitutional theory here, but the actual practice; in other words, first precisely how those decisions were made by Government in the first place, and in particular in the matter of this submission, how much control Parliament and their elected representatives therein were allowed to exert over those decisions. And so it is the intention here to show that Parliament has in fact never had the opportunity properly to debate or approve either this proposal before us today or any part of the whole network of major motorway trunk road proposals before the country at this moment.

And to this end, I wish first to refer you to the accompanying paper with relevant extracts from Hansard. With your permission, I will read it [See 'Extracts from Hansard' in Appendix 3, pp. 153–4].

The main points I wish briefly to make from it are:

1. That only on one occasion did any full debate take place with a motion put to the House and voted upon, namely on Mrs Barbara Castle's statement on 22 February 1967, proposing an expenditure of £1000m on inter-urban motorway construction;
2. That no vote was taken upon Dr John Woody's motion that an extended motorway plan should now be produced;
3. That only the briefest debate took place on the White Paper of May 1970, and that no vote was taken upon it;
4. That the all-important Road Strategy Paper of Mr Peter Walker of June 1971 was not even the subject of questions in the House; and that the Secretary of State subsequently deprived the House of any opportunity of debating and voting upon this all-important

transportation strategy document with its massive financial implications by refusing to publish it as a White Paper; and

5. That since then, a period of over four years, no full and proper debate (similar to that of 1967) has taken place and no vote has been taken, whereby Parliament might in any sense be seen to have given its approval to this massive commitment of national expenditure and resources; and that *in particular*, on neither of the two major issues subsequently affecting that commitment (namely, the likely availability of petroleum or any other motive energy at sufficiently low cost by the 1990s to justify further pursuance of the programme; and the revision of criteria and standards) has either debate or vote taken place.

To come forward a little in time, my clients wish me to draw your attention to a motion put forward in the House in the name of Mr Frank Hooley and signed by twenty-nine other M.P.s on 11 February 1975. This reads as follows:

> That this House is dismayed to see that central government expenditure in each year from 1974 to 1979 on motorways and trunk roads will exceed the combined central government subsidies to all forms of public transport by rail, bus and waterway (Cmnd 5897, Table 2.6, pp. 58–9) and calls for a Select Committee of the House to investigate as a matter of urgency the economic and environmental consequences of this policy.

And the point that my clients wish to make here, Sir, is that, of course, no such Select Committee has been set up. Once again, in other words, the House was denied the opportunity of any of that 'control' referred to in the Secretary of State's Statement in the matter of this proposal before you today.

1. *Mr Spearing*: Asked the Secretary of State if he will make a statement concerning approval of capital expenditure on road and rail improvements in S. E. England. *Mr Peter Walker*: 'No.' (Hansard, 21 July 1971)

2. *Mr Spearing*: 'I initiated a debate in this House on inland water transport. I put a number of questions to which I received no ministerial reply.' (Hansard, 1 May 1975)

Finally, I wish to come forward even further in time, with a few very brief extracts from the 1 May debate on the Public Expenditure Committee (Transport)'s Report:

Mr Huckfield: 'I wish that at least similar comparisons could be made between road and rail . . . Unfortunately, since we do not have the same figures, we cannot even make the necessary comparisions which would enable us to decide whether the Department is apportioning expenditure between road and rail on an optimal basis.

'Once again, the Report points up the need for a White Paper on transport policy. We have to know what the Government intend and what the Minister intends.

Many of us are still waiting for the emanation from Marsham Street of the White Paper which might set out the Government's intentions on transport. We have waited a long time.'

'Subject to the *control* of Parliament', Sir? My clients find themselves asking: Is there, perhaps, some Departmental joke-man responsible for this document, a sort of court jester. If so, Sir, they ask me to put to you very seriously that they do not take the joke. 'The control of Parliament.' They say that, from all the evidence available, Parliament is wallowing in just the same sea of *contrived* ignorance that overwhelms my clients. What *possible* control could exist under these circumstances?

And so, Sir, they submit, that until that control is restored, is a reality, this document is mendacious, that this enquiry is unconstitutional, and that accordingly you have no powers whatsoever to open it.

[The seventh submission, the subject of which is the corruption of the planning process in the Department, has been omitted as this is more comprehensively dealt with in the first and second submissions at the Archway inquiry (see pp. 65, 77).]

Appendix 3
The Winchester Submission

[Reference to the corruption of function of the Secretary of State's Department has been excluded from this submission as it is dealt with in greater detail in the first submission at the Archway inquiry (see p. 65).]

Submission: M3 Winchester Inquiry: 29 June 1976

Sir, in showing you that there are no grounds upon which you can open this inquiry scheduled under the Highways Acts 1959 and 1971 and the Acquisition of Land (Authorisation Procedure) Act of 1946, I require first to show you that the Line Order Inquiry into the M3 Motorway (Popham–Compton Section) Scheme 197 held here in Winchester during March and April 1971 was, for a number of reasons, invalid and defective. (And before I am told by Counsel for the Department that what I have to say on this matter is is irrelevant on the grounds that the Line Orders are already made or that a remedy existed for my clients in the Courts which, in that it was not taken up within a time specified by the Act, etc., renders their case invalid, I would request your patience, as I shall be touching on both these matters in due course.)

First

In the letter over the signature of Mr B. F. Edbrooke CSE 403/4/21/01 of 5 February 1973 and sent to all objectors (referred to colloquially as the 'decision letter') we find the following statement:

[Para. 67] Moreover, he [the Secretary of State] is convinced that there is an urgent need for this section of the M3 Motorway to be built as soon as possible in order to relieve the heavily congested and sub-standard Winchester By-pass, as well as the city of Winchester itself.

and in the document entitled DoE 'Statement to be Made by the Representative of the Secretary of State for the Environment', undated

[Para 1. 2. 4. (p. 3)] Since then, the Department has continually reviewed the situation having regard to the latest criteria for determining future traffic flows and road capacities and, in December 1975, in announcing the Secretary of State's decision to proceed with the scheme, the Minister stated that, 'in the light of the latest traffic data available taking account of the best estimates available of future population, car ownership, oil supply and economic trends, there is no doubt in my mind that this section of the M3 is still very necessary'.

Now, Sir, if we can consider those two statements together, what they mean is that the Secretary of State, taking into account the transport needs of this country (including traffic levels), within a coherent framework (which it is his duty to do of course) he has first fully understood them on a national and local level, and, within that comprehensive understanding of the nation's transport needs, he has concluded that, to relieve the heavily congested, sub-standard Winchester by-pass, etc., there is an urgent need for this section of the M3. These statements cannot have any other meaning. And so what I am required to ask you, Sir, is this: are those statements true, or to be more precise, can they or could they be true?

Well, Sir, in order to answer this question, I must now refer to the Departmental Paper *Transport Policy – A Consultation Document* (H.M.S.O., 1976).

1. First, to the foreword by the then-Secretary of State himself, and to the words therein: "Yet by common consent, we still lack a coherent national transport policy:"
2. Then to Section 4, Subsection 3, where we read: 'Transport policy will be properly co-ordinated at the national level only when there is a coherent framework.' (p. 22)
3. Section 13, Subsection 1 tells us: 'certain dominant themes have already emerged. By far the most important is the need to clarify the precise objectives of a national transport policy.' (p. 89)
4. And from the same section, Subsection 8 we learn that 'This approach' (an even-handed balanced approach to all aspects of

transport) 'cannot be pursued successfully without a proper framework for the co-ordination of transport policy at both national and local levels.' (p. 90)

And so, to summarise from these extracts, what do we find? Well, we still lack a coherent national transport policy; we don't have a coherent framework; we haven't yet even clarified the *objectives* of a national transport policy; nor do we even have a framework for the co-ordination of transport policy at either national or local levels.

Well, Sir, unless we are living in Alice's Looking Glass World, how, under these circumstances, in what possible manner, using what possible criteria, could the Secretary of State say in 1975 that there was no doubt in his mind that this section of the M3 is still very necessary or be convinced in 1973 that there is an urgent need for this section of the M3 motorway, etc.? Well, we are not living in Alice's Looking Glass World, but in the real world of 1976 and the plain fact of the matter is that if what the consultation document says is even half true, the Secretary of State could not now say that there is an urgent need for anything, let alone a six-lane motorway running around Winchester. He cannot say it now and he certainly could not say it in 1975 or 1973.

And so, Sir, as traffic does not exist *in vacuo*, but only in the context of all transportation modes, and as the Secretary of State's knowledge relating to the latter is of the value of nought, those statements are lies, and wholly mendacious. And what my clients say, therefore, is that they do not intend to have their whole environment put at risk, this unique English city endangered, those watermeadows concreted over, nor do those to whom it applies intend to see their property damaged for a scheme which is informed by what can only be described as lies and mendacious statements. And they say that, in that they are so, they invalidate the Secretary of State's decision on that Line Order Inquiry of 1971, which being so it makes a nonsense of the holding of this inquiry. Accordingly, Sir, they request you to abandon these proceedings forthwith, and sine die.

Second

I wish now to refer to the Highways Act 1959 Section 11, Subsection 6 where we read that the Secretary of State in proposing a scheme is required to take into account local and national planning.

Now, Sir, I don't wish to bore you with repetition, but my clients request me to ask you this question: If we lack a coherent national transport policy or a coherent framework; if we have not even clarified the *objectives* of a national transport policy; and if we don't have even a framework for the co-ordination of transport policy *at either national or local levels*, then how conceivably can the Secretary of State have taken into account the requirements of local and national planning?

The answer is that he cannot. And in that he cannot, he has failed to comply with that fundamental and most vital requirement under the Act. And this being so, my clients say, it invalidates the 1971 Line Order Inquiry and the Secretary of State's subsequent authorisation of the scheme. In that it does so, it invalidates this inquiry and accordingly again, my clients ask you to terminate these proceedings forthwith.

Third

A third matter I am required to raise before you relates to current Statements of Policy issued in the Secretary of State's name. These are issued to objectors at motorway and trunk road inquiries, and as these are all part of the national network, what statements are produced at one must apply to all. At the Aire Valley inquiry, now abandoned, and at the recent Haslingden By-pass inquiry this statement includes these words:

> This policy, and the size and shape of the roads programme in which it is given effect are matters for decision by the Government, subject to the control of Parliament.

Now, Sir, my clients require me to ask this question: given that this policy, etc. has been decided by Government (and by this we mean those very planners of highways in the DoE who act in the Secretary of State's name), has it been and is it in any sense as stated here 'subject to the control of Parliament'? The matter is one of the utmost importance. We are not here denying, and I hope that counsel for the Department will not waste our time in referring to this, that the 1959 Highways Act delegates powers to the Secretary of State's 'keep the road programme under review'. We are asking the very specific question: is the statement that the size and shape of the roads programme is subject to the control of Parliament true?

Well, Sir, if Parliament does in fact control the size and shape of the roads programme and the policy that gives effect to it, then this will be revealed in its debates. Well, debates on transport in the House of Commons are difficult to find. I have here a paper with the relevant extracts from Hansard and I wish briefly to make some points from it:

1. That only on one occasion in the last ten years did any full debate take place with a motion put to the House and voted upon (28 Feb 1967: expenditure of £1000m on inter-urban MWay construction);
2. That no vote was taken on Dr John Dunwoody's motion two days later that an extended motorway plan should now be produced;
3. That only the briefest debate took place on the White Paper of May 1970 with no vote;
4. That the all-important Road Strategy Paper of June 1971 was not even the subject of questions in the House and that the Secretary of State deprived the House of any opportunity of debating this all-important road strategy paper with all its massive financial implications by refusing to publish it as a White Paper; and
5. That since then, a period of over four years, with an exception which I will refer to in a moment, no full and proper debate has taken place and no vote has been taken whereby Parliament might *in any sense* be seen to have exerted any control over this policy with its massive commitment of national expenditure and resources.

On 11 February 1975 a motion was put forward in the House in the name of Mr Frank Hooley and signed by twenty-nine members. It read:

> That this House is dismayed to see that Central Government expenditure in each year from 1974 to 1979 on motorways and trunk roads will exceed the combined central Government subsidies to all forms of public transport . . . and calls for a Select Committee of the House to investigate as a matter of urgency the economic and environmental consequences of this policy.

No such select Committee has been set up of course. Once again, in other words, Parliament was denied the opportunity of any of that control referred to in the Statement.

But one debate did take place, and it is of singular importance. It was upon the House of Commons Select Committee Expenditure (Transport) Report (1 May 1975). Here are some extracts [all from

Mr Leslie Huckfield's contributions to the debate on the grounds that he commands a special position on the Government side in the matter of transport planning]:

When I see the detail into which the Report goes, such as the eleventh Recommendation, making the point that we still do not have the equipment, the yardsticks to compare various kinds of transport investment, I feel that the Report has hit upon one of the central matters lacking in present transport policies.

I wish that at least similar comparisons could be made between road and rail . . . Unfortunately, since we do not have the same figures, we cannot even make the necessary comparisons which would enable us to decide whether the Department is apportioning expenditure between road and rail on an optimal basis.

When we come to the inter-urban or inter-city investment which is made in transport infrastructure, we may come to the conclusion that one part of the Department does not know what the other part is doing.

Just as there appears to be no national co-ordination of our transport infrastructure or investment policies, there appear to be no local policies.

The Department does not even seem to be co-ordinating those parts of the transport industry for which it has complete responsibility.

Many of us are still waiting for the emanation from Marsham Street of the White Paper which might set out the Government's intentions on transport. We have waited a long time.

Once again the Report points up the need for a White Paper on transport policy. We have to know what the Government intend and what the Minister intends.

Now, Sir, we are not asking the question as to whether the Secretary of State (or those highway mandarins in Marsham Street who act in his name) can go on spending this nation's money at a rate of thousands of millions of pounds per year without recourse to Parliament. That is another question altogether, and one to which the country should very shortly address itself. The question we are asking is this: Is the statement 'This policy, and the size and shape of the roads programme in which it is given effect are . . . subject to the control of Parliament' true, or is that also a transparent untruth, a blatant lie? Well, Sir, with respect, my clients submit that the evidence of that 1 May 1975 debate is incontravertible. Control of a policy without knowledge of it is an impossibility in this world, and clearly knowledge of transportation matters does not lie with Parliament. But how could it be otherwise? We have seen from the Consultation Document that it does not lie with the Secretary of State either. In my client's submission, therefore, the line order inquiry of 1971, no less than this or any other road inquiry is based upon a lie – the lie that the policy is subject to the control of Parliament. And again, they say to you that they are not going to see their whole environment destroyed, their peace and quiet shattered, precious farmland put under concrete, and this whole area of country turned into a sprawling network of ever proliferating highways, and many of them with their private property irrevocably damaged while Parliament, and their elected representatives therein are denied any means of understanding and thus controlling the policy from which it is drawn.

And so, Sir, on this ground also, etc.

Fourth

Sir, my clients ask me now to make it clear to you that in the matter of this inquiry and, in so far as my investigations regarding the papers has revealed, in the matter of the 1971 Line Order Inquiry, the Secretary of State has failed to comply with a specific statement of his duties quoted to their noble Lordships in the Upper House by Baroness Stedman for the Department in answer to questions raised by Lords Molson, Avebury and Popplewell (25 Feb 1976).

The noble Baroness's words were:

> Before a trunk road inquiry opens, all objectors are given a statement explaining the general background of national transport policy and the main objectives of the Government's road building and improvement programme; and a further statement explains the purpose which the new road is intended to serve within the framework of the national system and goes into some detail about the route and its effect on the environment.

Now, Sir, you may or may not know, but this is precisely the requirement that the Council on Tribunals lays upon the Department in its letter to the Conservation Society of 18 January 1974 when, interpreting the Franks Committee's findings of 1957 they wrote that 'wherever possible some indication of the general policy relevant to the particular proposal under inquiry should be available before the inquiry'. (Correspondence from the Conservation Society to the Council on this matter eliciting this reply makes it quite clear that the words 'general policy relevant' do not mean and could not mean anything but *transport* as opposed to road transport policy, and transport policy is a complex of roads, railways, bus services, inland waterways, coastal shipping and pipelines.)

Well, Sir, those objectors whom I represent who were objectors to the Line Order Inquiry received a statement regarding the 'main objectives of the Government's road building and improvement programme', and something relating to the 'purpose which the new road is intended to serve', etc. But, Sir, they did not have any statement which could in any sense be regarded as, in the noble Baroness's words 'explaining the general background of national transport policy' or, in the words of the Council on Tribunals' letter, comprising the 'general policy relevant'.

Well, Sir, in that they did not have that statement, and in that the noble Baroness Stedman, speaking on behalf of the Department said that they should have had such a statement prior to the opening of that inquiry, it is my submission to you that that inquiry was *ab initio* improper in that objectors did not have adequate information upon which to formulate their objections. Accordingly on this ground also my clients require that, in that the original Line Order Inquiry was thus defective, this inquiry has no validity and that you abandon it *sine die* so that the Section 11 Inquiry can be properly convened and objectors allowed adequate information upon which to formulate their objections, with the statement explaining the 'general background of national policy' available to them 'before the inquiry opens'.

Fifth

I now require to refer again briefly to that ruling/interpretation of the Department's duties, given by the Council on Tribunals in their letter of January 1974. Therein two requirements are laid upon the Department:

1. As has already been mentioned, to provide information on 'general policy relevant' prior to the opening of the inquiry; and
2. to provide witnesses to answer questions on how the proposal fits into that general policy.

Now I wonder at this point if I could for a moment and by way of an illustration show what is meant by all this. Briefly, the Conservation

Society had said to the Council: We cannot understand motorways in isolation: as they affect all other modes of transport planning in a given area, we need information on all this before the inquiry in order to formulate a coherent objection, and we need the Department to put up witnesses, etc. And, as you see, the Council entirely conceded the point. But to the illustration:

Earlier this year British Rail NE Region issued a press release, as reported in the *Guardian*, which stated the following:

1. For the investment of around £½ million British Rail could achieve all the personal and freight movement (including thirteen million tons of freight annually) that was proposed to be achieved by the construction of the proposed M1/A1 link around Leeds at a sum of £100 million at 1974 prices.
2. This information, they stated, should be put upon the desk of the Inspector appointed to hold the motorway inquiry.

Now, Sir, it is not the mind-blowing fact of this revelation, and its proof of the extent to which every motorway section is nothing less than a national economic disaster, which I wish to impress upon you at the moment – but rather the fact that this is *precisely* the sort of thing that the Conservation Society and the Council on Tribunals had in mind. The Department is responsible for *transport* planning – as the Consultation Document [and the setting up of the Baldwin Steering Committee] so belatedly acknowledges. *It* should therefore, in proposing the M1/A1 link be required to show in what way it fits in with rail potential for the area, etc. How otherwise can objectors come to any proper conclusions on the matter of the need for the motorway?

Now let us see what happened when the Conservation Society presented an objection at the M65 inquiry in Burnley in March of 1974. As that motorway is proposed to run parallel to a railway line, the Conservation Society required to know what the Government's plans were for the railways of that region. I acted for the Conservation Society on this occasion, and when I asked this question, Counsel for the Department [Mr Fay Q.C.] got up and said the words: 'I certainly do not intend to put up a witness to answer questions about railways. This is a road inquiry.'

Now, Sir, to return to the M3 Line Order Inquiry. In the first place, an examination of all the papers reveals that objectors at that inquiry received no statement complying with the first requirement. And in the second, it is clear from a close reading of the inspector's report of that inquiry that no witnesses were available from the Department who could *in any sense* be said to comply with the requirement of the Council on Tribunals that they should answer questions on 'how the proposal fits in with that general policy'. They were, as I recall, merely technical witnesses concerned with engineering and traffic, route landscaping and noise.

In that they did not do so, they can only be said to have failed to comply with the second of the requirements of the Council's interpretation of the Franks Committee Report of 1957. And, therefore, in that the Department have thereby failed to comply with both requirements, my clients say that the information available to them, both before and during the inquiry was totally inadequate in order to formulate any proper objection. Accordingly, the inquiry took the form it did, being no more than a list of attempts of one group or another to shift the line onto one set of property or another. In that it did so, my clients say again that the Line Order Inquiry be re-convened with information on 'general policy relevant' available to them before the inquiry opens, and with witnesses available to explain that policy during the inquiry.

Sixth

Sir, if we refer again to the Consultation Document, and to Section 9, Subsection 23 (p. 72) therein we read as follows:

> The procedures, both statutory and non-statutory, for deciding whether or not a trunk road should be built, have given rise to a lot of complaint. As has already been announced, further changes are planned. The aim will be to make the present arrangements more acceptable and to improve the presentation and clarity of the information provided to objectors . . .

And if we refer to the Departmental Press Notice of 20 January of this year we note that the Minister for Planning and Local Government stated in Parliament the following: 'I am pleased to say that the Council on Tribunals have assured me of their willingness to join with us in re-examining the adequacy of our procedures.'

Now, Sir, the acknowledgement by the Secretary of State of the inadequacy of his present procedures applying to motorway and trunk road inquiries is both implicit and explicit from the above extracts. Accordingly, my clients put it plainly to you, Sir, that they see no reason why they should not benefit from this procedural review. *They* do not find the present arrangements acceptable; *they* require an improvement in the presentation and clarity of the information provided to them (and in that so much is at stake they see no reason whatsoever why they should not have it); but most of all, they require the benefit of the Council on Tribunals' involvement in the matter of 'the adequacy of procedures'. This, for reasons which I have just made clear, they see as crucial. And, Sir, they do not take any suggestion from the Department that this will lead to delay. This for two reasons. First, delay in decision is clearly what is required if the Consultation Document means anything at all. And second, the delay is due entirely to the Department's refusal to take account of legitimate complaints until years of frustration resulted in the regrettable uproar associated with the Airedale inquiry. I would remind you that the urgent need for a procedural review was implicit in the Council's letter to the Conservation Society as long ago as January 1974.

What my clients say to you, therefore, very plainly, is that they do not intend to have their properties damaged, this unique city threatened and their environment destroyed until the Line Order Inquiry where this applies has been re-convened and held in a manner avoiding those procedural inadequacies referred to and implied in the Minister's statement.

Seventh

May I deal now with a matter already mentioned, but implicit in all that I have been saying this last ten minutes, namely with the all-important matter of an objectors' right to object to the *need* for a given road scheme?

The origin of the acceptance of this elementary right under the Act was before yourself, Sir, in June 1973 at the M42 inquiry at Bromsgrove, and has since been clearly acknowledged by the Department in its Observations (Para. 9) on the First Expenditure Committee Report 1974 that 'the Government agrees that the need for a road scheme may appropriately be challenged at a public inquiry, provided that matters of policy are not called into question'. And it is true to say that prior to that, objectors were, quite contrary to the Act, denied this right.

I will take as an example of this the M16 (A13–A12) Section Inquiry which was taking place concurrently with the M42 Bromsgrove inquiry. There, Mr Fay Q.C. (again) stated these words: 'Ringway 3 is part of a strategic network which has already been approved by Parliament. We cannot challenge this here. This inquiry is concerned with alignment and landscaping.' And following this, the Inspector, in speaking to Counsel for the Department used the words: 'I have not had any directive' and said that he would make a note of the fact that objectors might have been inhibited in presenting their cases. He followed this with the statement that it was not unknown for an inquiry to have to be re-opened.

Well, Sir, to come again to the M3 1971 Section 11 Inquiry what do we find? Well, at once we find a restriction imposed. If you will refer, Sir, to what is colloquially referred to as the M3 S.E.R.C.U. Line Order Statement Part 1, Para 1.2. Here we read the unequivocal words: 'The proposals to be considered at this Inquiry relate only to the alignment of the new road. When a line has been fixed by the Secretary of State, proposals for interchanges as and for alterations to the existing road system will be published in due course . . . etc.'

Well, Sir, I do not need to emphasise the restrictive nature of this statement. Anyone reading it would at once realise that if he wished to object on the grounds that the road should not be built at all because it wasn't needed, formed no part of a national transport strategy or was, for one reason or another contrary to the national interest, any such person would at once realise that he would be denied the right to do so. He would assume, and in my submission be correct in assuming, that should he attempt to do so, he would meet with a rebuttal akin to Mr Fay's at the 1973 M16 inquiry. And that, of course, is contrary to the Act, which does not so restrict objections on the grounds of need.

My clients, or some of them, here today will put it very forcibly to you, Sir, that restrictions *were* in fact placed upon them, and that they were denied the right to raise these matters. The whole tenor of the inquiry in this matter was eminently expressed in the Inspector's reference to the closing address by Mr R. H. McCall for the Winchester City Council. Para. 65.5 reads: 'There have been many individual objections but they have all adopted a negative attitude – they did not want the motorway so they closed their eyes to the problems which were staring them in the face.'

One wonders which is the greater, the sheer arrogance or the sheer ignorance of this statement in the face of the admissions to be found in the Consultation Document. But that is not, of course, the point I make here. It is the fact that those arrogant words, spoken towards the conclusion of the inquiry, clearly illustrate the contempt received by any objection which might remotely be conceived as having sought to show that there was no need for the motorway.

And if that were not conclusive, under the heading in the Inspector's Report issued to objectors, *Findings of Fact, Observations and Conclusions*, Para. 69.3, we read the dismissive words:

> The opposition to the scheme from individuals and representative organisations was mostly of a negative nature. They did not want a motorway at any cost but, with the exception of the Upper Itchen Valley Society, they generally appeared to refuse to face up to the problems that exist at the present time and which will rapidly deteriorate as traffic volumes increase year by year.

And we are now at the very nub of this issue. Perhaps there is and was no need for this motorway. Certainly, as mentioned earlier, the Consultation Document reveals that we are in no position whatever to say whether there is or there isn't. But from this extraordinary statement in his report the inspector reveals two things: first, that he simply was not interested in any objections which were not concerned with the line of route, and second, the very real weakness of his report to the Secretary of State as a direct consequence. The assumption that, just because the Department and its agents in the County Councils state that there is a need, a need necessarily exists was never a valid one, and the Consultation Document has now gone to prove that incontravertibly. Parliament did not assume the totality of wisdom to repose in the Department – otherwise it would have placed restrictive conditions upon the nature of objections [in the Act]. Need *is* to be questioned. Need *should* be questioned. These *seem* and *are* urgent requirements of a non-totalitarian society. But need was not questioned at the 1971 Line Order Inquiry. It was not questioned because the right to do so was denied, and any attempt to do so was treated with contempt. And in that the right to question need was denied by Para. 1.2 of the S.E.R.C.U. Line Order Statement, that Line Order Inquiry was invalid when it was held in March–April 1971, and requires now to be held again. That is my clients' submission, and it is their submission accordingly that they do not see themselves today in any way subject to the requirements of the Acquisition of Land (Authorisation Procedure) Act of 1946. Or to put it in plainer terms, they do not intend to see any land and properties compulsorily purchased so long as the necessary and preliminary statutory procedures have not been complied with by the Secretary of State.

* * *

And may I come now, Sir, to what I am sure Counsel for the Department might otherwise be quick to point out when he is on his feet, namely to the matter of my clients having had a remedy in the courts relating to the validity of the Line Order Inquiry and the possible fact that, as they failed to avail themselves of it, they now have no case. The Ex Parte Ostler appeal court judgement, fully reported in *The Times* Law Report of 17 March of this year, refers. Well, Sir, in order to show you that the 16 March Appeal Court Judgement cannot be said to apply to my clients, I require to make five brief points:

May I first say, Sir, that we are here concerned with matters which go beyond the narrow confinements of statute law. After all, there are other forms of law. There is administrative and tribunal law, interpreted by that body set up by Parliament to do so, the Council on Tribunals. There are the ancient customs of our unwritten constitution, one of the fundamental principles of which is that moneys raised by taxation shall be subjected to scrutiny in the spending thereof by our elected representatives in the Commons. There is the law of natural justice, with which, Sir, I do solemnly submit, we are going to have to concern ourselves with this morning. And it is interesting to note that this was the basis of a House of Lords decision upholding a recent appeal on a Compulsory Purchase Order Inquiry of 1973 – on the grounds that it 'had been invalidated for want of natural justice'.

Point one Many of them had no advice. They were without any legal representation. And on this matter it is relevant for me to draw your attention to comments by Mr Geoffrey Pattie, M.P. for Chertsey-Walton, in the *Surrey Herald* of 29 April 1976. He was, significantly enough, commenting upon the Line Order Inquiry into the Chertsey–Wisley Section of the M25 (an inquiry at which I was personally present, and upon the conduct of which I shall be required to comment when the Compulsory Purchase Order inquiry is convened). Therein the M.P. suggests that there should be a mandatory requirement:

that all objectors should be represented on inquiries by Counsel unless they specifically ask not to be. The corollary of this is that legal aid should be available to those objectors who otherwise could not afford to be represented. This particular reform on planning inquiries is long overdue, and in my view, essential in order to equalise the advantage or disadvantage to each party.

Mr Pattie, M.P. is quite right, and it was precisely because of this deprivation of adequate legal advice that my clients found themselves unable to follow the complex legal requirements available to them in the remedy according to the Act.

Point two How were they to know the *basis* upon which they could deny the validity of that Line Order Inquiry? At the time of the issue of the Decision Letter of 3 February 1973 the Department was still denying objectors their rights under the Act; indeed the M.R.C.U. sought to deny Mrs Barbara Maude her right to contest the need for the M42 motorway right up to the date when, as mentioned earlier, Mr Harold Marnham, Q.C., showed this for the illegality that it was.

Now, Sir, those great mandarins in the Department who so far have decided all these matters amongst themselves, and who delegate Under Secretaries to sign Decision Letters on behalf of themselves (calling themselves the Secretary of State of course), these men of great power know and knew of that section. They know that the 1959 Act gave their inspectors no powers to deny all objections except those to the line of route. And this being so, Sir, it is my clients' submission that, in that the decision letter did not include a statement to the effect that, in denying objectors their right to object on the grounds of need, the inquiry might have inhibited them (to use Mr Rolphe's well-chosen phrase), the Decision Letter is itself mendacious and thus denied my clients the essential information upon which they could have decided whether or not they had been so inhibited, and thus whether or not to apply to the High Court within the six weeks specified by the Act. In other words, Sir, to put it more plainly, when the Secretary of State denies persons the knowledge necessary to make a decision to apply the statutory remedy, he can hardly thereafter say that he will not hear their plea because they failed to apply the said remedy.

Point three Much of the basis, as you will readily appreciate, Sir, upon which my clients claim Departmental impropriety does not relate to the conduct of the Inquiry itself, but with the lies contained in the Under Secretary's letter, and that lie in the Secretary of State's Statement of Policy relating to Parliamentary control. It doesn't need much knowledge in law, Sir, to appreciate that these are not matters which the Court would concern itself with, and in that they are not, the requirement to apply to the court cannot apply in this instance.

Point four You cannot yourself, Sir, decide upon these matters: you cannot make decisions upon legal matters. And furthermore, of course, you may *listen to*, but you cannot *act upon* the advice on these complex and contentious matters of Counsel for the Department. Of course, his views will be radically different from mine, but he is, and do not let us forget it, no less a party to this dispute than I am. This leads me to a further adverse comment upon the Chertsey–Wisley Section (M25) Inquiry by Mr Pattie, M.P.: As reported, he stated: 'He [the inspector] is not a legally qualified person and this is the first flaw in the present system. In a quasi-judicial setting where groups of objectors are represented by counsel, it is imperative that legal assessors be appointed to advise inspectors on points of law'.

If we look back to the predicament of Mr Rolphe, who at one point was reported saying, 'I have had no directive', we can see how inadequately there were the interests of objectors maintained, but that is not the point that I seek to make here. It is your present incapacity in these matters which here concerns us. There is but one remedy for you, Sir, and that is to adjourn this inquiry now and refer all these matters back to the Secretary of State for his consideration.

May we ascertain what your powers are in this matter, Sir, by referring to the 1971 Act Section 54?

Well, Sir, there we read:

> Where proceedings . . . are taken after the confirmation or making by the Secretary of State of an order or scheme . . . *the Secretary of State* may disregard for the purposes of the said Schedule 1 any objection to the compulsory purchase order or draft thereof, as the case may be, which in his opinion amounts in substance to an objection to that order or scheme.

Well, Sir, my clients are objecting to that scheme of 1971 on the grounds of need and on the grounds that they were prevented from presenting such objections. And the words I now wish to draw your attention to in what I have just read are all in my line 2. You will note, Sir, that the words are 'the Secretary of State': the act does not say 'the Secretary of State or the person appointed'. So clearly, Sir, you are not empowered to make a decision on this matter. You must at once adjourn this inquiry and place the matter before the Secretary of State, who (and this is where I wish to draw your attention to a second point in the quoted section) *may* disregard (or otherwise) any such objection. (the word, Sir, is not 'shall').

Point five Sir, there are two categories into which some of my clients fall for which failure to seek the remedy as required by the 1959 Act could not possibly deny them a hearing in this matter of reconvening the Line Order Inquiry.

In the first place, some of my clients now objecting to this Side Roads and Compulsory Purchase Order Inquiry were not objectors at the 1971 Line Order Inquiry. One of the reasons for this, in my submission, is the total inadequacy of the statutory notice and the Department's repeated failure to comply with the 1959 Act First Schedule, Part II, Para. 7. As you must know, Sir, the Conservation Society is in dispute with the Department in this matter, with the support of Mr George Dobry Q.C. and now Mr David Widdicombe Q.C., whose official opinion obtained by the society opens with the words: 'I agree with the view Mr George Dobry Q.C. put forward at the M16 inquiry in that the notice published by the DoE was defective in that it did not sufficiently state "the general effect" of the road scheme.'

Well, Sir, this is not the place or the time to pursue this matter, but sufficient to say that many of my clients have now become very well aware of the general effect of this particular road scheme and now wish to object to the side roads and compulsory purchase orders, and are entitled to do so.

But, as they were not objectors to the Line Order Inquiry, and as they now, having examined its conduct, are profoundly convinced that it was improper and requires to be held again, they can hardly be denied this on the grounds that they have failed to apply for a writ in the High Court within a given period of six weeks. In this case, of course, they are no different from Mr Ostler, but here we are again concerned with natural justice. Indeed, were Mr Ostler to appeal to the House of Lords, who knows but that he might be upheld 'for want of natural justice'?

Much the same can be said for those of my clients who were not resident here at the time of the Line Order Inquiry, but who now, examining the conduct of that inquiry deny its propriety and legality. The fact that they were not here, either in 1971 or 1973, means that they could not possibly be expected to apply the prescribed remedy. The fact that they are here now, however, and find that their whole environment is threatened by a scheme, the inquiry into which was, in their submission, improperly conducted, fully entitles them to demand that that inquiry be reconvened, and that the following requirements be complied with therein: Parliamentary approval for the scheme be shown to have been obtained: compliance with both of the Council on Tribunals' requirements complied with: that it be convened only after the completion of the procedural review; that the right to object to the *need* for the scheme be fully accepted and advertised beforehand, and that the findings of the Baldwin Steering Committee be awaited and applied. That is their entitlement, and upon that they stand.

To Show that no Parliamentary Approval has ever been Granted for the Present Motorway/Trunk Road/Lorry Route Programme

Extracts from Hansard

1. 22 Feb 1967. Minister of Transport (Castle)'s statement (Cmnd Paper 3057 of 20 August 1966). On M/Ways it proposed expenditure of £1000m on inter-urban construction.
   A debate took place. House approved by 320 to 233.
2. 24 Feb 1967. Private Member's motion (*Dr John Dunwoody*): 'That in the opinion of this House . . . an extended motorway plan should now be produced.' Commenced 12.09hrs; only 13 Members took part; 'It being 4 o'clock the debate stood adjourned.'
   No vote was taken.
3. 7 June 1967. Question (*Sir Clive Bossom*): 'Can the P.U.S. say when we are to get the 2700 miles of M/Way that the County Surveyors' Society maintain are essential?' *Mr Swingler*: 'We are all set to achieve 1000 m by early 1970s.'
4. 28 Nov 1968. Question (*Sir W. Bromley Davenport*): When is the Minister to 'announce the M/Ways on a national basis which will form the network which is to supplement the *present proposed system*?' *Mr Marsh*: 'I hope to publish proposals for a future

inter-urban highways strategy as soon as possible in the New Year.'

5. 26 March 1969. Green Paper published: *Roads for the Future – a New Inter-Urban Plan.*
6. 8 Dec 1969. In answer to a written question on the Green Paper from Sir C. Bossom. *Mr Mulley*: 'I have received replies from 265 organisations . . . I hope to announce decisions on the shape of the future inter-urban strategic road network early next year.'
7. 28 Jan 1970; 24 Feb 1970; 28 April 1970. Three requests for White Paper (M.P.s Dodds, Parker and Heseltine (2)).
8. 27 May 1970. Mr Mulley introduced his White Paper on Inter-Urban Road Strategy for England. Debate of approx 30 minutes; 14 Members spoke. No vote was taken.
9. 23 June 1971. In answer to a question by Mr John Hannan, the Secretary of State (Mr Peter Walker) announced the Inter-Urban Road Strategy Paper. There were no questions, no debate and no vote.
10. 6 July 1971. Question: Mr Nigel Spearing asked the Secretary of State if he will publish his proposals as a White Paper. *Mr Peter Walker*. 'No.'
11. 7 March 1973. Question: Mr R. C. Mitchell asked the Secretary of State if he will publish a White Paper on his long-term transport policy. *Mr Peyton*: 'Not yet, Sir.'
12. 5 Dec 1973. In reply to a question from Mr Jay, *Mr Rippon*: 'The Government are soon to publish a White Paper on Transport Policy. That Policy will reflect the potential contribution of the M/Way programme and the likely availability of oil supplies.' This White Paper has never been produced.
13. 17 June 1974. In reply to a question from Mr Alec Jones, Mr Mulley announced future plans for roads, including a crucial revision of standards. No debate and no vote.

Appendix 4
Professor Terence Morris's Archway Submission

The Al London–Edinburgh–Thurso Trunk Road, Archway Road (Haringey) Improvement (Winchester Road – Great North Road) (Trunk Road and Slip Roads) Order 197

The Al London–Edinburgh–Thurso Trunk Road, Archway Road (Haringey) (Winchester Road – Great North Side Roads) Order 197

Submission to the Inspector Appointed to Re-open a Public Local Inquiry on 15 September 1976

The clients whom I represent require me to make certain representations to you at the outset of these proceedings, and in particular have requested me to ensure that they are made prior to the taking of appearances, to reading of any statements by you, or any other formal proceeding whatsoever, for this reason

THAT

precedent shows that when such representations are made after such formalities both inspectors and the Department of the Environment have seen fit to regard inquiries as having been opened and are prepared only to consider such matters as part of the general substance of objections to proposed schemes and orders

WHEREAS

the representations I am required to make to you affect the propriety of the continuation of the proceedings of this public local inquiry and must therefore in both logic and justice to my clients be taken before any other submission relating to the substance of the outstanding parts of the draft orders published in connection with this scheme.

I propose to outline in brief the nature of these matters in order that you may have a clear indication of what is involved and then deal with each of them at greater length.

I am required categorically to demand of you that you adjourn all proceedings in connection with these schemes *sine die* on the following grounds:

1. That the public local inquiry held in 1973 and 1974 did not properly consider the need for this road scheme and objectors were not given adequate opportunity to dispute the assumption of the Department of the Environment that such construction was either necessary or desirable;
2. That a precedent now exists whereby, even when the line of a proposed new road has been confirmed by the Secretary of State and a public local inquiry has been arranged to consider objections to proposals for side roads only, an inspector may permit objectors to raise again objections which are objections to the scheme in principle, notwithstanding an official statement to objectors that the Secretary of State may disregard such objections should he so choose, and that those objectors to this scheme whose objections were to the scheme in its entirety and not merely to specific lines or constructional details should now be given the opportunity adequately to prepare such objections taking into account factors which have emerged since 1974, and that these proceedings should be adjourned to permit them to do so;
3. That the Department of the Environment's *Transport Consultation Document*, (H.M.S.O., 1976) reveals the total absence of any national or local data upon which major decisions affecting transport policy can be made;
4. That as the Document recognises, the procedures of inquiries into road proposals have given rise to much complaint, and the conduct of the inquiry into these proposals in 1973–4 is not exempted, and that since the Secretary of State has stated that a re-examination of the procedures has been commenced in co-operation with the Council on Tribunals it would be wholly improper and contrary to natural justice that these proceedings should re-open prior to the completion of this review and its adequate discussion by Parliament;
5. That these Orders, both those already made and those still in draft,

have been or will be made under the powers of the Secretary of State under the Highways Acts, 1959 to 1971 and other enabling powers, in that they are being made without being credibly subject to the control of Parliament and in the light of the fact that they are intended to be made against the wishes of a very large number of local residents which the Secretary of State seeks to disregard, are a constitutional abuse which ought immediately to cease until such time as adequate Parliamentary control has been established over the executive acts of the Secretary of State in relation to road construction programmes specifically and transport policy generally.

These, then, are in brief the grounds upon which I am asked to submit that it would be wholly wrong for you to attempt to continue these proceedings. In order that it may be made plain that these are matters of the utmost gravity, and not merely an expression of dissatisfaction with the conclusions of the Report of the 1973–4 proceedings by the inspector, Mr F. H. Clinch, I propose to turn to each of them in detail.

1. I am sure that it will quickly be argued by counsel for the Department that the need for this particular road scheme, far from being ignored at the earlier proceedings, was ventilated to the full, not only by the Department but by local residents who considered that the environmental conditions prevailing on the Archway Road were now so intolerable that something must urgently be done to provide some relief.

   Now, Sir, let me say at the outset, that none but the most eccentric could deny that conditions on the Archway Road are appalling and that the volume of traffic, often slow moving, has contributed significantly to the decline of environmental standards. But the remedy for this state of affairs does not necessarily lie in creating a new road, or in making an existing road wider by the demolition of existing properties. Merely to make roads larger and wider because they are over-crowded is to pay the modern equivalent of Danegeld. Every improvement in an urban road, which is an approach road likely to be used by commuters or commercial operators, attracts the marginal user who would otherwise have been deterred. If one examines the so-called 'improvement' schemes in London that have been embarked upon in the last two decades one finds that they have contributed nothing save to the increasing commitment of the population to private at the expense of public transport. In spite, for example, of vast traffic schemes in the area west of Hyde Park Corner towards Hammersmith and the M4, including enormous road works along the Cromwell Road, conditions are now worse than ever. On the other side of the coin, the cost of all such urban road schemes is extremely high in human terms. I would suggest, Sir, that you see for yourself some of the utter desolation that has been created by the construction of new road schemes, not merely in areas of great poverty such as North Kensington where slums have been made worse slums by elevated roadways, but in areas of middle-class settlement, where, in order that roads might be made wider, homes patiently bought over the years are now set in environmental conditions that can only be described as squalid.

Now Sir, I do not know what view you will take of such statements. The Department takes the view, as is well known, that environmental factors cannot be accounted for within its systems of cost-benefit analysis. From a reading of the Report by Mr Clinch it is clear that he was not inclined to take very seriously objections which he considered as 'emotional'. But the point is this: to regard as 'emotional' the proposition that the construction of new urban roads, far from improving the environment by means of traffic management merely results in the increase of urban squalor and human misery from noise, vibration and pollution from exhaust gases, quite apart from the loss of property values, is conveniently to side-step the issue of the responsibility of both Government ministers and their highway planners. It may be that in a totalitarian society the rights and feelings of individuals can be cast aside, but we do not, as yet, live in such a society. To exclude environmental factors from any analysis of cost-benefit is to deny the validity of the claims of ordinary people who live and work in an area to a certain standard of decency and to confirm the interests of those who choose increasingly to exploit the use of the motor vehicle as paramount.

I do not know whether you have read the Report made by Mr Clinch, but whether you have or not, I would urge you to look closely at the general burden of the case made by objectors who saw this scheme as doing nothing but worsen the conditions of an environment already seriously impaired by heavy traffic, and consider it in the light of the general emphasis of the case made by the Department. You will see that the assumption that there will be an inevitable growth in road traffic which must be accomodated by ever-continuing road development is dominant. The whole of the inspector's Report is dominated by the acceptance of these assumptions, and a variant of utilitarian philosophy which argues that the few must sometimes suffer in the interests of the many.

Now it is not just that those whom I represent had no adequate opportunity to establish the case that the rights of the few, to be spared even further disruption in their lives, are at least as important as those of the many who are frustrated in their motor vehicles by each moment of delay in the Archway Road as they strive to reach northwards to Edinburgh – or even to Thurso. It is that there was no clear and reasoned case indicating why the interests of road traffic were such that they should take precedence over all other forms of transport. Moreover, by the time that the inquiry had begun a shift in policy by the Greater London Council had occurred in the direction of actually placing restraints on traffic within the London area by limiting new road building and in particular for this area, the proposed extension to the Ml. That this change of direction should not be to the liking of the Department can come as no surprise, yet it is abundantly clear that at no time was the suggestion that traffic in the Archway Road area ought to be *restrained* given any serious consideration.

The real question, Sir, is whether such a construction can be justified by showing that the need for it is so great and the benefits so obvious that all other costs can be set at naught. This can only be done if the view is taken that the demands of road transport must take precedence over all other considerations. It is in this sense that the need for this scheme was not properly considered let alone established.

2. In the Statement of Reasons sent to objectors on 2 August last [1976] by the Department it was made plain that the object of these proceedings was to re-examine the outstanding parts of the draft orders which had been made in part only following a recommendation of the inspector. This was confirmed by a letter to objectors from a Mr B. B. Bradford dated 25 August 1976. It is quite clear that the Department, from this Statement, regards the whole scheme as *fait accompli* save for certain details relating to access roads. This view is not shared by my clients who seek to have the case for the scheme as a whole properly examined in the light not only of all social and environmental considerations

relevant to the inhabitants of the area, but taking into account changes in economic factors affecting road transport.'

In thus arguing, my clients draw to your attention the fact that at the Public Local Inquiry held in Winchester in June and July of this year into draft orders relating to side roads in connection with the proposed section of the M3 Motorway from Popham to Compton, the inspector, Major General R. C. A. Edge, agreed, *inter alia*, to the proposition that the case for the line of the motorway could be considered, notwithstanding that it had already been confirmed by the Secretary of State after a Public Local Inquiry and that objectors had been warned that the Secretary of State might disregard such objections to the side roads proposals as he considered to be objections to the whole motorway scheme in general. It was there clearly established that the operative words derived from the Statute are '*may* disregard' and are not in the imperative form of '*shall*'. At the original Public Local Inquiry at Winchester the opportunity to discuss the need for the motorway as opposed to a line had been denied. My clients are of the opinion that although a great deal was spoken about the need to improve conditions in the Archway Road, such a concept of need was predicated upon the view that the only solution to problems of road traffic congestion is the construction of more and larger roads; was based upon the assumption that the demands of road users must take precedence over all other demands, including the demand to be able to enjoy a reasonably peaceful environment free from the spectre of ever growing volumes of traffic, and that the whole proceedings were dominated by a disregard for the feelings and wishes of those whose homes are threatened by these proposals.

Given that objectors have been given to understand that the purpose of these proceedings is to inquire into the final details of certain slip roads it would be unreasonable, were you now to agree that general objections might be heard, that they should be expected to present a case against the scheme as a whole here and now, not least because all the data which were cited at the last inquiry are now out of date, in some cases so out of date as to be not merely worthless but actively misleading. For that reason they ask that these proceedings be adjourned *sine die*.

3. I refer now to the paper *Transport Policy: A Consultation Document* in the foreword of which the then Secretary of State

wrote: 'Yet, by common consent, we still lack a coherent national transport policy . . .' Section 4 (3) includes the words: 'Transport policy will be properly co-ordinated at the national level only when there is a coherent framework . . .'

Section 13 (1) states: ' . . . certain dominant themes have already emerged. By far the most important is the need to clarify the precise objectives of a national transport policy . . .' and in Section 13 (8) we read that a balanced approach to all aspects of transport ' . . . cannot be pursued successfully without a proper framework for the co-ordination of transport policy at both local and national levels'.

It is abundantly clear from this document that there is no coherent transport policy. There is neither a coherent framework for co-ordinating policy nor even a clear set of objectives. Now Sir, what all this must mean, and indeed can only mean, is that the Department of the Environment, which by its actions is so patently committed to the construction of more and more roads and which has in fact enlarged its own structure in order to do so more effectively, is determined to press on with the implementation of policies which will assure the total dominance of road transport over other forms of transport, and in the process the dominance of private transport over public transport. It is clearly determined to do this notwithstanding the absence of a coherent national transport policy to which fact the Secretary of State has made admission. Like a belligerent before an impending truce it seems as if it wishes to establish irreversible tactical advantages over its opponents before its position is immobilised by a moratorium on further public expenditure on the roads programme.

The state of affairs revealed by the Consultative Document is serious enough, but the position is exacerbated by the fact that there are serious inhibitions upon any attempt to discuss local road proposals in the light of either regional or national transport policy, assuming that transport policy means all forms of transport and not merely road transport. It has been consistently maintained that local inquiries cannot represent such a forum. Yet it is equally clear that neither within the council chambers of local authorities, nor within Parliament itself is there any significant opportunity for the mass of the electorate, through their representatives, to make any impact upon the course of government action in regard to the roads programme. Indeed, the increase of scale in local government operations and the nature of the links between the Department of the Environment's highway engineers and their counterparts within the local authorities is such as to encourage the growth of a labyrinthine bureaucratic structure which stems the flow of information to the public and its elected representatives. The result is that the ordinary public is not merely confused but ignored. Meanwhile those who are committed to a philosophy of increasing the volume of road transport at the expense of other forms have very free rein to act as they please.

Given that there is no coherent national transport policy the assignment of such importance to a road intended to link London with Thurso – or more locally to provide access to the Ml and other parts of the motorway network – is absurd. Those whom I represent maintain that until there is some recognisable form of coherent transport policy, taking into account the needs of Greater London in relation to the rest of the country, the continuation of these proceedings which will facilitate the construction of the proposed scheme ought to be adjourned.

4. I come now to the question of the procedures of inquiries. You cannot fail to be aware, Sir, that there is grave discontent about the way in which such inquiries are conducted, including the manner in which the Secretary of State may, if he so chooses, simply set aside findings which are not favourable to the policies of his Department. There is also the question of whether it is either proper or desirable for an interested party, such as the Secretary of State, not only to appoint the Inspector, but to control through his junior staff the very arrangements of the local inquiry itself. I may tell you that at the earlier inquiry into this proposal, certain private objectors were limited in their opportunity to cross examine witnesses for statutory bodies appearing at the inquiry.

Now on 26 July last in the adjournment debate in the House of Commons on the M3 motorway inquiry, the Under-Secretary of State Mr Marks, in seeking to answer the charge that the Secretary of State acts as 'judge and jury in his own case' stated: 'Such complaints are misconceived. Trunk road inquiries and the decisions the Secretary of State has to make after considering Inspectors' Reports are administrative, not judicial' (Hansard, col. 218). That, Sir, is not an answer, but merely a compounding of justification for a practice which is in itself objectionable, for it is in the administrative process that seeks to make decisions in terms of the interests of road builders and road users at the expense of those who object that injustice is done. Small wonder that Mr Marks denies that the process is judicial. It is a process which denies the most elementary natural justice to the ordinary citizen who is, individually, almost powerless in the face of the bureaucratic power ranged against him. Although those who have protested vigorously at certain road inquiries have been accused of being undemocratic by Ministers within the Department those ministers well know that there are many things wrong. I would commend you the first leading article in *The Times* newspaper for 30 August which refers to the 'present widespread public distrust of road inquiries'. That, if anything is an understatement. The fact that procedural matters have been referred to the Council on Tribunals is an encouraging sign, but it would be a more convincing indication of Ministerial good faith if there were to be a suspension of all inquiries, pending not merely the outcome of that review, but the adequate discussion of its findings by Parliament. That there is no such moratorium proposed suggests most strongly that the Department of the Environment is determined to press on with its programmes of road construction taking the best advantage that it can from the totally unsatisfactory state of affairs now characterising road inquiries. You Sir, are however in the position of being able to suspend this particular inquiry pending the outcome of the review by the simple expedient of adjourning it *sine die*, which is what is now requested of you.

5. I come now to the final ground upon which my clients seek an adjournment *sine die* which may be termed the ground of constitutional abuse. That is not a term which is used in any way lightly, but is intended to make the point that we have now reached such a stage in these affairs nationally that can be no longer tolerated. It is no longer a question of whether a mile or so of a road between London and Thurso or anywhere is constructed to run this way or that, to demolish this or that property but a question of how monies raised by taxation are spent. It has taken several hundred years for the principle that monies so raised should be spent under the proper scrutiny of Parliament, yet with regard to the roads programme there is no evidence that Parliament has any effective control at all. Since 1970 only one major debate has taken place in the House of Commons on the

Report of the Select Committee on Expenditure (Transport) in May 1975. But apart from revealing that there is no coherent national transport policy – which the subsequent Consultative Document admits – there is no evidence that control of the purse strings resides anywhere save within the Department of the Environment, subject to the internal constraints on government expenditure.

But the situation is worse than merely that created by a lack of effective Parliamentary control; there is reason to believe that there is comparatively little Ministerial control either. That is to say the Secretary of State, and those junior ministers subordinate to him, are wholly dependent for their information upon the permanent officials of the Department of the Environment who advise them. It is a melancholy fact that when individual members of the public write to ministers the replies they receive are sent by permanent officials, sometimes so junior in status that it is clear that the communication has never even reached the desk of one who may speak to the minister, let alone the minister himself. Reporting upon the comments by the Council for the Protection of Rural England on the Consultative Document, *The Times* for 30 August notes the Council as saying that: 'Parliament has been grievously remiss in allowing successive governments to maintain a rolling programme of massive road construction without scrutiny of its justification.'

Ministers come and go in Marsham Street but the permanent officials are to all intents and purposes just that. The fact that a succession of Ministers should have had no effect in halting the roads programme until such times as the guidelines for a national transport policy were clear, indicates just how far the staff of the Department are able to determine the course of events by the manipulation of executive powers.

Now, Sir, it is the case that the Department has pursued its ends by means that are wholly unacceptable in a democratic society. By the proposal to build a piece of road here and another piece there, the alignment of one section forecloses options on the next. Archway Road could scarcely be a better example. When plans for one section are made public there is a reluctance to discuss what other plans there might be, even though they are clearly in existence. By suggesting that a road proposal will improve the lot of one set of citizens – albeit at the cost of another – the Department succeeds in setting one group against another. Indeed, there is nothing that it likes better than seeing one group arguing the case for dumping unwanted traffic and several hundred thousand tons of concrete on another.

It may be suggested that to object to a road proposal in terms such as these is irrelevant. Indeed, it might be argued that the matters I have raised are political matters, and have nothing to do with a road proposal. Nothing could be further from the truth; on the contrary, it is precisely the aseptic non-political context of administration that makes it possible for a government department to act against the interests of the local inhabitants of this area. They have seen how, present their arguments with what skill and moderation they might at the 1973 inquiry, their views were set aside. For now they realise that even if the inspector had recommended against the project altogether, the Secretary of State, or rather those permanent officials who lurk in anonymity behind his name, would have over-ridden the advice. They have also seen how, when the patience of reasonable people has been tried beyond endurance,the Department of the Environment has seen fit to employ the police to remove objectors from public inquiries. Has no one inside the Department asked himself whether the use of the police in such circumstances is not evocative of the practices of the totalitarian state? Has no one asked how it comes about that ordinary people who are law abiding in their everyday lives are prepared if need be to risk criminal prosecution rather than be silent when they believe that they are being denied what are their constitutional rights and freedoms under law? For that is a final point that must not escape notice. It may be said by counsel for the Department that what has been done by administrative law cannot be readily undone. It is part of the argument of my clients that there is a conflict between administrative law and procedure and their rights not only under natural law but under the unwritten constitution of this country. It is not denied that ministers must be able to exercise certain powers which they must delegate to permanent officials. But this is not the same as suggesting that the use of executive powers should grow to the point where Parliamentary control is no longer credible as a restraint over expenditure specifically or policy generally. You, Sir, have it within your power today to take notice of these real and important concerns and to act. The earth will not cease to turn upon its axis, nor will the town of Thurso waste away from want of communication if this so-called road 'improvement' is not immediately built. But by bringing these proceedings to a close you cannot fail to impress upon the Secretary of State that there are graver, more fundamental issues to be attended to. Nor can you fail to impress upon Parliament the urgent need to re-gain control over an important area of public expenditure and by so doing, halt the insidious drift towards bureaucratic totalitarianism. You may do so, quite simply, by adjourning these proceedings *sine die*.